Dangerous Liaisons?

PUBLISHING FOR THE WORLD
125 Years

THE JOHNS HOPKINS UNIVERSITY PRESS

Syntheses in Ecology and Evolution
Samuel M. Scheiner, Series Editor

Dangerous Liaisons?

When Cultivated Plants Mate with Their Wild Relatives

Norman C. Ellstrand

The Johns Hopkins University Press
Baltimore and London

© 2003 The Johns Hopkins University Press
All rights reserved. Published 2003
Printed in the United States of America on acid-free paper

Johns Hopkins Paperbacks edition, 2005
9 8 7 6 5 4 3 2 1

The Johns Hopkins University Press
2715 North Charles Street
Baltimore, Maryland 21218-4363
www.press.jhu.edu

*The Library of Congress has cataloged the hardcover edition of this
book as follows:*

Ellstrand, Norman Carl.
  Dangerous liaisons? : when cultivated plants mate with their
wild relatives / Norman C. Ellstrand.
     p.   cm.— (Syntheses in ecology and evolution)
  Includes bibliographical references (p.   ) and index.
  ISBN 0-8018-7405-X (hardcover : alk. paper)
     1. Transgenic plants. 2. Transgenic plants—Risk assessment.
  3. Pollination. 4. Plant diversity conservation. I. Title. II. Series.
  SB123.57.E44 2003
  631.5'233—dc21

                                        2002156768

ISBN 0-8018-8190-0 (paperback)

A catalog record for this book is available from the
British Library.

To Mom, Dad, Tracy, and Nathan

Contents

Figures and Tables

Figures

Tables

Preface

The pauses between the notes—ah, that is where the art resides!
—ARTUR SCHNABEL, PIANIST

The most exciting research occurs where the frontiers of two fields intersect.
—LESLIE GOTTLIEB, PLANT GENETICIST

"Feed the world!" "Frankenfood!" "Evolution revolution!" "Killer tomatoes!"

The loud and often emotional public discussion of the positive and negative impacts of genetically engineered plants is hard to ignore. At one extreme is self-righteous panic, and at the other, smug optimism. Between those extremes is the simple fact that any new technology has the potential for solving problems, but no technology is without the potential for some negative impacts. While the genetic engineering controversy plays out in the media and on the streets, dozens of scientists have been at work to collect data to examine the potential risks of these engineered, transgenic crops.

To anticipate the possible risks of transgenic plants, plant ecologists and population geneticists have examined the problems associated with traditionally improved crops. The most widely discussed have been (1) crop-to-wild hybridization resulting in the evolution of increased weediness in wild relatives, (2) evolution of pests resistant to the new strategies for their control, and (3) impacts on nontarget species in associated ecosystems (such as the unintentional poisoning of beneficial insects) (e.g., Snow and Moran-Palma 1997, Hails 2000, Wolfenbarger and Phifer 2000).

In the 1980s a few plant scientists (e.g., Colwell et al. 1985, Goodman and Newell 1985) first recognized that the first on the list, crops mating with their wild relatives, could prove to be an avenue for the escape of engineered genes, and those escaped genes might create environmental problems. Two Calgene scientists stated the concern clearly: "The sexual transfer of genes to weedy species to create a more persistent weed is probably the greatest environmental risk of planting a new variety of crop species" (Goodman and Newell 1985). Unaware of that publication and other prior work, I presented a similar idea as a three-minute tag-on to a half-hour plant gene-flow talk at a professional meeting in the summer of 1987. Reluctantly accepting an invitation to write up that idea (Ellstrand 1988), I unknowingly took my first step in the direction of what is now a major theme in my research program.

In the late 1980s the prevailing view was that hybridization between crops and their wild relatives occurred infrequently, if at all. Furthermore, if hybrids did occur, it was anticipated that they would be sterile or nearly sterile, unable to pass on their genes to future generations in the wild. That view was supported by the belief that the evolutionary pathway of domestication leads to reproductive isolation, including hybrid sterility. The primary support for that belief was the fact that plant breeders are often challenged in their attempts to transfer desirable genes from wild plants to domesticates by hybridization, both by inability to create hybrids and by the sterility of the few hybrids they are able to create.

But a plant breeder's hand-crosses in the greenhouse are not the same as natural pollen exchange in the field. Thus, my research group set out to measure spontaneous hybridization between pairs of related crops and weeds: first, cultivated and wild radish (Klinger et al. 1991, 1992) and then grain sorghum and johnsongrass (Arriola and Ellstrand 1996). We found that hybridization occurred at greater distances and higher rates than anticipated by the prevailing view. We also grew the hybrids under field conditions to see if they were as fertile as their parents. In both systems the hybrids were as fit as their parents (Arriola and Ellstrand 1987, Klinger and Ellstrand 1994); in the case of the radishes the hybrids were even more fit. Although we didn't use genetically engineered plants for our experiments, it was clear that, at least for these species, engineered genes could easily escape into and persist in wild populations.

But when I gave seminars on the results of these experiments, I was met by questions: Are these results general for most domesticated plants? Maybe there's something special about radish and sorghum. Do we really need to worry about engineered genes? After all, if hybridization with wild relatives creates problems, wouldn't those problems be well known for traditionally improved crops? These were good questions, but ones that required research time devoted to searching libraries and databases, with fact-checking by phone, mail, and Internet. I started what I had planned to be a four-month literature review for my 1993 sabbatical at Uppsala University but that ended up as a multiyear odyssey whose fruit is the book in your hands.

And it has been an exciting odyssey. Like nuggets in a riverbed, the information I needed was spread thinly and sporadically over more than a dozen fields of study. Making sense of what I found was the challenge and the reward. Interestingly, this research question marked one of the chasms that separate academic fields. In this case those fields are the evolutionary biology of natural plant populations and crop genetics. The hybridization process joins those fields. The last few decades have seen growing interest in the evolutionary significance of hybridization in both plants and animals (e.g., Arnold 1997). But most evolutionists remain focused on natural systems, despite the fact that agricultural systems present spectacular and well-researched examples of evolutionary change. In the same way, a crop geneticist tends to focus on his or her individual crop, even though research on other crops and even natural systems may have direct implications for that person's work. Like each of the famous Indian blind men inspecting the elephant, each biologist perceives only a tiny part of biological reality (Ellstrand 2000). The depth and breadth of our ignorance is humbling. But it's also a simple fact of life. Biology is immensely complex. In a world of ever accelerating information, we scientists are forced to pick and choose what facts we are going to assimilate, ever narrowing our focus just to keep up.

And even more exciting than discovering the gaps was the opportunity to bridge them. This project has provided the opportunity to pull together theory and data from fields as diverse as plant systematics and evolution, biotechnology, agronomy and horticulture, theoretical population genetics, ethnobotany, plant breeding, weed science, conservation biology, and risk assessment. The quotations that kick off this preface ring true. Focusing on a topic that falls between fields is indeed exciting—and, because interfaces have no boundaries, a bit overwhelming.

Therefore, I have tried to structure this book to be tractable to a broad audience. As much as possible, each chapter is created as a freestanding unit. Readers well versed in plant evolutionary genetics can probably skip or skim chapters 2 through 4. Practical scientists, such as policy makers or conservation managers, may prefer to jump around from one chapter to another. For many, particularly those readers not yet comfortable with plant evolution and population genetics, the linear approach may be best. Each chapter tends to build on the prior ones.

Part 1 of the book ("Foreplay") provides the necessary background for readers who are not plant population geneticists. It examines the general issues of spontaneous (natural) hybridization and introgression—and their potential consequences:

— Chapter 1 introduces the three-part real-life hybridization drama, "The Case of the Bolting Beets," the story of the newly evolved weed beets of Europe.
— Chapter 2 sets the stage for the main thread of the book by explaining the basic terminology of hybridization for the nonspecialist, outlining the scope of the book, and describing the romantic intrigues of plants that can bring genes with cultivated ancestry into the wild.
— Chapter 3 addresses the question, How often does plant hybridization occur in nature? Natural hybridization, even between species, is in fact a common feature of plants. This chapter provides an overview of spontaneous hybridization among plant species in the wild to demonstrate what we might expect for the cases involving spontaneous mating between crops and their wild relatives.
— Chapter 4 moves on to the next logical question, So what? Does hybridization make any difference? The chapter examines the evolutionary consequences of hybridization and its applied implications. Population-genetic theory predicts that the flow of genes from one population into another by hybridization can play a potent role in the evolution of the population receiving those genes, sometimes even overwhelming the effects of natural selection. Chapter 4 reviews the theoretical consequences of gene flow when interacting with the other evolutionary mechanisms and makes some predictions about what might happen when domesticated plants mate with their wild relatives.

Part 2 ("Caught in the Act") pulls together what is known about spontaneous hybridization between domesticated plants and their wild relatives, focusing on the evidence for their natural hybridization:

— It is important to consider the kinds of detective work that support the contention that such hybridization actually occurs. What evidence is used to determine whether plants naturally hybridize? In Chapter 5 the types of evidence for hybridization most frequently used are described and evaluated, setting the stage for asking whether domesticated plants spontaneously hybridize with their wild relatives.

— Chapter 6 returns to "The Case of the Bolting Beets" and builds on chapter 5. That chapter is a case study, showing how scientists amassed evidence to demonstrate that Europe's weed beets were the result of hybridization between sugar beets and wild beets.

— Beets are but one example. Do the most important products of plant domestication hybridize with their wild relatives? Chapter 7 presents in-depth treatments of the situation for the world's twenty-five most important food crops, demonstrating that most of them naturally hybridize with one or more wild relatives somewhere in the world.

— But is hybridization with wild relatives a general feature of domesticated plants? Chapter 8 addresses that question.

Part 3 ("Dangerous Liaisons?") turns to the significance of these liaisons:

— Are the predictions from population-genetic theory borne out as observed consequences of domesticated-wild hybridization and introgression? Chapter 9 compares the expectations outlined in chapter 4 with the known consequences of gene flow from crops into wild populations.

— The final three chapters consider the "special" case of genetically engineered crops. Chapter 10 presents the last chapter of "The Case of the Bolting Beets," focusing on whether gene flow from genetically engineered beets to wild beets might create special problems.

— Are engineered plants going to behave differently from the products of conventional plant breeding? Chapter 11 takes up where chapter

10 leaves off, considering the issue of gene flow from genetically engineered crops more broadly.

— Gene flow from crops to their wild relatives can have both desirable and undesirable consequences. If humans are mindful of the consequences of gene flow from domesticated plants into wild populations, then they have the opportunity to make decisions about what to do about it. Chapter 12 concludes the book with a discussion of options for managing and monitoring gene flow.

Acknowledgments

No substantial project is ever the product of a single person working in isolation. This one could not have succeeded without the generous support I received from people and institutions during its development. Much of that support started even before I knew that what I was doing would lead to a book. The groundwork that got me going was financially supported by a Fulbright Fellowship to Sweden, a National Science Foundation Mid-Career Fellowship, and grants from the United States Department of Agriculture (#94-33120-0372 and #00-33120-9801), UC MEXUS, and the Swedish Forestry and Agricultural Research Council.

But intellectual and emotional support is more important than money. I am grateful to all of those who have taught me and made me think about hybridization. Loren Rieseberg became the first of these teachers during a sabbatical in his lab in 1992. He is still my favorite hybridization guru, promptly and patiently answering my questions.

Honor Prentice and Jim Hancock are two more bodhisattvas who co-authored articles (Ellstrand et al. 1999, 2002) that served as a point of departure for this bigger effort. Helena Parrow helped with the initial compilation of case studies for that article; they have now evolved into the treatments for crops in chapter 7. I am particularly grateful to Honor and Jim for their friendship and their encouragement to proceed with the book. Syndallas Baughman and Anne Rankin prereviewed that article and made dozens of wonderful suggestions that would have improved it greatly. Regrettably, making those changes would have doubled the length of an already-too-long manuscript. Fortunately, their suggestions fit comfortably into this book.

Members of my lab working on crop-to-wild hybridization projects have been a source of thoughtful inspiration: Janet Clegg, Terrie Klinger, Paul Arriola, Detlef Bartsch, Lesley Blancas, Subray Hegde, and Roberto Guadagnuolo have all brought their talents to bear on studies exploring whether domesticated plants mate with wild relatives and, if so, what it all means.

And what it means for the escape of engineered genes has been a theme that our lab has puzzled over for more than a decade. Individuals calling for stringent regulation of genetically engineered crops have shared their concerns and discussed the issue with me at length. Beth Burrows of the Edmonds Institute, Jane Rissler and Marty Mellon of the Union of Concerned Scientists, and Becky Goldburg of the Environmental Defense Fund have reminded me that science doesn't stop at the lab door. Likewise, biotechnology industry scientists concerned with creating the best possible sustainable products are equally concerned about making sure that transgene escape doesn't create any problems. I have enjoyed teaching and being taught by Tom Nickson and Mike Horak from Monsanto as well as Donna Mitten from Aventis. Those who are caught in the middle of the issue, the regulators, perhaps have the hardest charge of all. Karen Hokanson, Craig Roseland, John Turner, and James White of USDA-APHIS have repeatedly opened my eyes to the challenges to policy-making scientists in the trenches.

Despite a stack of manuscripts competing for my attention, my book proposal was written and submitted during my sabbatical at the University of California at Irvine in 1998–1999. I am grateful to Francisco Ayala for serving as my sabbatical mentor and providing me with the resources to get the job done. I appreciate the support from my UCI friends who were willing to chat when I was in need of coffee, chocolate, or ethnic food: in particular, Francisco Ayala, Jim Bever, Diane Campbell, Steve Frank, Ann Sakai, Art Weis, and Steve Weller. That sabbatical year also gave me an opportunity to renew my friendship with my major professor Don Levin at the University of Texas. I found that he is just as good at teaching how to write a book as he was at teaching how to write a paper. Staff Research Associate Janet Clegg, the heart and hands of my lab at the University of California at Riverside, kept things going at home while I was trying to "think great thoughts" at Irvine. Knowing I could trust Janet to get the job done made that Irvine sabbatical my most academically productive year.

During my sabbatical Ginger Berman helped me visualize what I wanted to create with this book while persuading me to choose the Johns Hopkins University Press as my publisher. After Ginger left JHUP, editors Sam Scheiner, Sam Schmidt, Trevor Lipscombe, and Jim Jordan were willing to give me attention when I needed it and to leave me alone when I didn't. Special thanks too to Mary Yates for her thorough polishing of the manuscript. Most important, I never felt that anyone at JHUP took me for granted.

I asked a very large number of folks to comment on specific sections of the manuscript. The following scientists prereviewed the case studies in chapters 6 and 7: J. Antonovics (rice), P. Arriola (maize and sorghum), D. Bartsch (sugar beet), L. Blancas (maize), K. Breure (oil palm), A. H. D. Brown (barley and soybean), E. Buckler (maize), T. T. Chang (rice), H. Corley (oil palm), J. M. J. de Wet (finger millet), J. Doebley (maize and sorghum), J. Doyle (soybean), J. Eckenwalder (sweet potato), J. Ehlers (cowpea), P. Garcia (oats), D. Garvin (beans), P. Gepts (beans), R. Guadagnuolo (wheat), A. Hall (cowpea), W. Hanna (pearl millet), H. Harries (coconut), S. Hegde (groundnut), K. Hilu (finger millet), T. Holtsford (maize), T. Hymowitz (soybean), F. Ibarra-Perez (beans), R. Jørgensen (barley and rapeseed), P. Keim (soybean), J. Kohn (rice), D. LaBonte (sweet potato), H. Morishima (rice), P. Morrell (sorghum), N. Nassar (cassava), H. Oka (rice), R. Palmer (soybean), M. Perez de la Vega (oats), B. Reisch (grape), L. Rieseberg (sunflower), Y. Sano (rice), K. Schertz (sorghum), A. Scholz (finger millet), U. Schuch (coffee), B. Sobral (sugarcane), D. Spooner (potato), H. Suh (rice), R. Tayyar (chickpea), M. Van Slageren (wheat), R. von Bothmer (barley), J. G. Waines (beans, rye, and wheat), N. Weeden (grape), and J. Wendel (cotton). Arlee Montalvo and Subray Hegde made key suggestions that refined chapter 4. Peter Morrell's firm, but gentle, preview of chapter 5 helped to sharpen its scientific detail and reduce my bias toward and against certain techniques. Jodie Holt made sure that I didn't write anything off base with regard to herbicide resistance. Allison Snow, Sam Scheiner, and Linda Newstrom-Lloyd heroically read and critiqued the entire first-draft manuscript end to end. I am grateful for all of the reviewers' thoughtful responses and the time and energy they spent teaching me. The book benefited considerably from their suggestions.

The following individuals supplied key pieces of hard-to-find information, ranging from ideas for quotations to kick off the chapters to the value of the world's sugar crop: C. Aquadro, M. Culver, T. D'Hertefeldt, C. Funk, P. Gepts, V. Gibeault, L. Hall, S. Kafka, G. A. Sakar, and L. Wolfe.

I am grateful to Beverly Ellstrand for creating "Environmental Fallout," the beautiful painting on the cover of this book and the visualization of genetic pollution by gene flow. Thanks also for Janet Clegg's skillfully created photos for figures 3.1 and 5.2. Dr. Michael Hamilton of the University of California James Reserve helped to locate the oak leaves displayed in figure 3.1 at that reserve.

Finally, I never realized how important it was to have people who believe in you—in this case, people who believed that I could create something as daunting as a book. How enabling it is to have folks you care about say, "Of course you can write a book!" So, special thanks to my favorite colleague and spouse, Tracy Kahn, my pop culture consultant and son, Nathan Ellstrand, and my parents, Edwin and Beverly Ellstrand. I cannot thank you all enough.

Part I / Foreplay

CHAPTER 1

The Case of the Bolting Beets

Part I. Long-Distance Romance

> Birds do it, bees do it,
> And in France even beets do it.   —WITH APOLOGIES TO COLE PORTER

A warm, golden afternoon in the south of France is a perfect time and place for romance—especially for a wild beet.

The air slowly rises in the heat of the day, and the result is natural air-conditioning. Soon after sunrise a cool, steady breeze from the Mediterranean Sea moves inland, filling the void left by the heated air. The sea breeze carries dust from the beaches, tiny salt crystals spun from sea spray, and pollen from the coastal plants.

For much of that pollen the journey is a lonely one. Most coastal plant species hug the sea strand. Pollen blown inland from those species is doomed to death just as surely as that blown out to sea. But pollen of a few species may find a mate kilometers from home.

Sea beets *(Beta vulgaris* ssp. *maritima)* are one example. Although they are tightly restricted to the coasts of northern Europe, sea beets have a more relaxed distribution along the Mediterranean, where they occur both on the French coast and in human-disturbed habitats in the inland valleys of southwestern France (Letschert 1993, Boudry et al. 1994a). Sea beets are the progenitors of all cultivated beets *(Beta vulgaris* ssp. *vulgaris):* those for human consumption (e.g., table beets and chards), those to be eaten by animals (e.g., mangolds), and those to be processed for that important industrial biochemical, crystalline sugar (sugar beets) (Ford-Lloyd 1995).

The morning breeze from the Mediterranean picks up wild sea beet pollen from plants on the coast. Because beets are mostly wind-pollinated, that pollen can remain viable over considerable distances, even many kilometers. The pollen may find a mate among its cousins growing along the roadsides of the inland valleys. Alternatively, the wind may carry both coastal and inland sea beet pollen to a somewhat more distantly related mate, sugar beet.

Sugar beet seed is an important agricultural product in the inland valleys of the south of France. Indeed, much of northern Europe's sugar beet industry depends on that seed. Modern commercial sugar beet varieties are hybrids between two different inbred lines of the same crop. That hybrid seed has the benefits of hybrid vigor; that is, interline hybrids typically have very high yields relative to their inbred parents. To create hybrid sugar beet seeds, rows of a male-sterile (i.e., female-only) line are planted in a field alternating with rows of a male-fertile line (Kernick 1961). In isolation of all other beet pollen, seeds set by the male-sterile plants are interline hybrids, with hybrid vigor and combining the best of the two parental types (Ford-Lloyd 1995).

The word "isolation" is key. European sugar beet breeders recommend that sugar beet seed production fields be isolated by one kilometer from the nearest flowering beet to obtain 99.99 percent pure seed (Barocka 1985). A field of a few hundred table beets left to flower by a bankrupt truck farmer could produce enough pollen to contaminate a nearby sugar beet seed production field such that a good fraction of the seed produced would be commercially worthless sugar beet × table beet hybrids. That is why the south of France has been chosen as a sugar beet seed production area: because it is far from the major beet production areas of Europe.

And indeed, France's sugar beet seed production fields are many kilometers from the tens of thousands of sea beets growing along the Golfe du Lion and at least one kilometer from those growing in the inland valleys. It turns out, however, that those fields are not so far from the sea beets as to frustrate long-distance romance. The stray pollen grains borne by the late-morning wind have turned out to have considerably more significance than a stolen kiss. By the time those illicit romances were identified, they had produced international consequences.

Rosettes of green leaves arranged in neat rows, stretching for as far as the eye can see under the bright blue skies of the north of France—sugar beet fields at their best can be a breathtakingly beautiful sight. But by the mid-1970s northwestern Europe's sugar beet fields had become pocked with tall and spindly plants, spangled with seeds. These skeletal weeds were a metaphor for lost profits and economic damage to come. The sugar beet growers of northern Europe received a rude shock. Too many of their beets were flowering prematurely. In the jargon of sugar beet folks, they were "bolting."

Well-behaved sugar beets don't bolt. They behave as biennials. That is, during their first season of growth they put up a rosette of leaves and put down a substantial succulent root. In the fall the plant is harvested. The leaves are cut off for animal food. The homely root, containing about 20 percent sugar, is processed to produce crystalline sugar. If beets are left in the field, by accident or intentionally for seed production, the cold of winter triggers a change in their lifestyle. They bolt during the second season; that is, these "vernalized" plants send up a flowering stalk, depleting both their leaves, which wither, and their swollen root, which becomes woody and useless. They set seed and die (Ford-Lloyd 1995).

Renegade beets live fast and die young. That is, they behave as annuals. After a few months as rosettes they bolt, drawing all of their resources from leaves and root to nourish their flowering stalk and seeds. Every field of beets has a few individuals that are off on their timing and accidentally bolt. Typically, no more than one in a million sugar beets, or even fewer, exhibit such deviant behavior (D. Bartsch, personal communication).

Imagine the surprise of the sugar beet growers of northern Europe, from France and Britain to Germany and beyond, when their sugar beet fields were suddenly infested with hundreds and thousands of bolters. Obviously, a bolter deprives a farmer of a useful plant—in terms of both withered leaves and woody root. But a bolting beet is more obnoxious than you might think. Bolters rise above well-behaved crop plants, stealing sunshine and depressing the yield of their neighbors (Longden 1989). And the woody roots of bolters take their toll on farm machinery. They may damage machinery at the processing plants as well. Everyone is willing to put up with the odd bolter, but when they become too common the field might not even be worth harvesting. And the frequency of these bolters was increasing.

Generally, sugar beet bolters do not increase in frequency for three reasons. First, your standard bolter might have some physiological quirk, but generally not a genetic one. Most frequently an unexpectedly late cold spell stimulates bolting in beets that were planted a bit too early. Thus, if the seed these vernalized bolters drop back into the field germinates the next time a farmer plants sugar beet seed, the children of these plants will grow up to be biennials. Therefore, the farmer who practices short rotations between sugar beet crops doesn't have much of a problem if vernalized bolters were present in recent crops, because if the seeds from those plants germinate at the same time as the crop, they usually grow up to be well-behaved beets. Second, bolters are typically so infrequent in a field that they have trouble finding a mate. Beets are generally self-incompatible plants (Fryxell 1957). Self-incompatible individuals are incapable of pollinating themselves and require pollen from another plant. So if bolting beets occur sporadically in a field, here and there, surrounded by hundreds of thousands of chaste biennials, they don't set much seed. Third, the prudent farmer knows that bolters should be eradicated and sends out teams of workers to scour the fields and remove the odd bolter by hand.

But the new bolters of Europe were different on two counts. First, they appeared at a relatively high frequency, typically about eight out of one hundred thousand plants (Mücher et al. 2000), one hundred times more frequently than expected. Mate-finding opportunities increased with the number of bolters. Therefore, they were able to set plenty of seed. A single bolter can produce as many as twenty thousand seeds (D. Bartsch, personal communication). Second, their bolting wasn't a physiological quirk; it was genetically determined (Boudry et al. 1994a). Most seeds set by the newly deemed "weed beets" germinated to become bolters. European farmers soon realized that bolters were increasing dramatically with each year of sugar beet production as seed from the bolters created a seed bank in the soil of their fields.

The plague of weed beets spread rapidly. By 1981 they infested 1.2 million hectares of northwestern Europe's sugar beet fields, an area roughly the size of Belgium (Longden 1993). In the ensuing decades they caused millions of dollars' worth of problems for Europe's sugar industry.

Although sugar beet represents a multibillion-dollar industry and is Europe's primary source of sugar, it is declining in importance as one of the world's major crops. In 1970 sugar beet was the world's twenty-second most important crop in terms of area planted; by 2000 that ranking had dropped to twenty-eighth (FAOSTAT Statistics Database, http://apps.fao.org/). The same time period saw a 21 percent decrease in area planted while most important crops were enjoying an expansion. Interestingly, sugarcane steadily increased in area planted. Weed beets are a component of that change. Crystalline sugar is a product whose global value exceeds $9 billion per year (S. Kaffka, personal communication). Sugar is more than a sweetener for desserts and beverages. It is an ingredient in nonedible products such as transparent soaps; furthermore, it is a chemical precursor for other biochemicals such as alcohols and resins. As demand for sugar increased worldwide and sugar from beets became increasing costly, sugarcane obtained an advantage. The problem isn't going away. As I write, some European sugar beet fields are growing more weed beets than sugar beets, as many as eighty thousand bolters per hectare (Mücher et al. 2000).

And control of weed beets is a headache, costly and difficult (Longden 1993). Weed management must include strategies that reduce both the number of weed beets in the field and the number of seeds in the seed bank, the population of ungerminated seeds in the soil. Because weed beets are the same species as sugar beets, their seedlings and rosettes look the same, until they bolt. Any herbicide that can be used safely on sugar beets has no effect on their weedy cousins. Cultivation or selective application of herbicides can remove those weeds that sprout between planted rows, but only a few management practices work to control weed beets that are intermingled with the crop. If broad-spectrum herbicides are applied prior to planting the crop, then most germinating weeds are killed. Sometimes a tractor-drawn wick, soaked in herbicide and held above the crop rosettes, is used to selectively kill the taller bolters that it contacts. These are expensive and imprecise solutions. Long periods of crop rotation, that is, growing crops other than sugar beets for several years until the weed's seed bank is depleted, is more effective because weed beets are competitively inferior to most of northern Europe's crops. But short periods of rotation do little to deplete the seed bank, because only a fraction of the seeds germinate in any given year. An effective, but time-consuming, method is to wait for the weeds to bolt and then kill them one at a time through hand-weeding. These complex management programs work to reduce weed beets to acceptable levels, but they aren't cheap or easy (Longden 1993).

Beyond management were bigger questions: Why were the bolters increasing so rapidly? Where did these plants come from? Clearly, weed control wasn't going to do the job if every batch of sugar beet seed was contaminated with bolters. One theory was that they were sugar beets gone bad, atavistic mutants that had reverted to the annual form. Another theory was that these plants were annual wild beets, the subspecies native to the coasts of Europe and the evolutionary ancestors of all cultivated beets. Perhaps seeds from sea beets had contaminated sugar beet seed. Or perhaps the wild beets had mutated to a form that preferred the good life of the cultivated field to the rigors of the coast. A third theory—and one eventually proven to be true—was that the bolters were hybrids between sugar beets and their wild relatives, the products of long-distance romance. Solving that mystery involved some substantial genetic detective work detailed in the second part of this story.

Eager to continue with "The Case of the Bolting Beets"? Pick up the story in chapter 6. Otherwise, proceed to chapter 2.

Hybridization and Gene Flow, an Introduction

HYBRID, n. A pooled issue.   —AMBROSE BIERCE, 1911

Hybrids fascinate. They are thrillingly exotic and at the same time a bit scary. The fantastic creatures of the ancients such as the griffin, with the head and wings of an eagle and the body of a lion, embody the tension between these two aspects of hybrids. Mixing supposedly discrete entities can evoke positive images (consider "hybrid vigor," "introducing new blood," and "fusion cuisine") or negative ones (contrast the emotional impact of "half-breed," "mutt," or "miscegenation").

Hybrids are an affront to those who feel that the world should be an orderly place. Basic biology textbooks teach that species are defined as groups of individuals that do not interbreed with other groups. Hybrids, if mentioned at all, are usually treated as biological oddities, best suited for the sideshow. The "Dog-Headed Boy" notwithstanding, hybridization is real. Hybridization happens, by human intervention as well as spontaneously in nature between individuals of populations that have not yet diverged to the point of absolute reproductive isolation. As detailed in chapter 3, hybridization is not so common that all species blur together into a blended soup, nor is it so rare that every individual in the world can be categorized to a given species.

The present chapter provides a point of departure for the rest of the book: outlining its scope, defining some basic terms for the nonspecialist, and describing how natural hybridization can bring domesticated genes into the wild. This book's focus is hybridization that occurs when domesticated plants naturally mate with their wild relatives—and the consequences of that hybridization. It concerns itself with the evolutionary biologist's definition of hybridization (Arnold 1997, Harrison 1990), that is, crosses between individuals from genetically distinct populations. In most cases the crosses discussed in this book are between species or subspecies. Hybridization frequently leads to introgression, the incorporation of genes specific to one taxon into one or more populations of another (adapted from the definition of Richards 1997).

I define "wild" plants as those that are essentially free-living, that is, those that grow and reproduce without being deliberately planted and cared for. The wild relatives of crops might be plants with no domesticated ancestors in their genealogy, but they might also be the descendants of crops long ago gone wild or even be plants with a hybrid history involving both wild and domesticated ancestors. Note that the term "feral," indicating the recent spontaneous descendants of an intentionally planted crop, also falls within this definition of a wild plant, provided that the feral plants are more or less self-sustaining. In practice, it is often difficult to identify whether spontaneous crop plants are ephemeral volunteers, emerging over a number of years from a seed bank, or self-sustaining feral populations (cf. Lutman 1993, Pessel et al. 2001). (I define "weeds" as plants that interfere with human objectives, but a different definition is used in fig. 8.1.)

Much of this book is devoted to how hybridization and introgression can lead to evolutionary change. Evolutionary change depends on the activities of the four evolutionary forces: natural selection, mutation, genetic drift, and gene flow. Whether between species, subspecies, races, or simply genetically distinct populations, hybridization and introgression are subsets of gene flow, defined as "the incorporation of genes into the gene pool of one population from one or more other populations" (Futuyma 1998). As explained in later chapters, the evolutionary consequences of hybridization can be profound.

This book is also about hybridization at certain conceptual levels. It is about the hybridization of two traditionally discrete research areas: natural plant population genetics and agricultural plant population genetics. While the two fields share the same fundamentals, few scientists have straddled both at once. Some of the great plant evolutionists—Allard, Anderson, Baker, de

Wet, Harlan, Heiser, Levin, Stebbins—made substantial contributions to our understanding of the population genetics of both natural and agricultural species. Nonetheless, even these scientists (with the exception of de Wet and Harlan) typically chose to study either natural populations or agricultural populations in any given project. Understanding the consequences of crop genes in natural populations requires thinking about both natural and agricultural populations at the same time.

The book addresses another conceptual hybridization: the mating of plant population genetics and public policy associated with the possible environmental impacts of genetically engineered (transgenic) plants. Those two areas were joined in a shotgun marriage in the 1980s, after the realization that hybridization could act as an avenue for the escape of engineered crop genes (transgenes) into natural plant populations (e.g., Colwell et al. 1985, Goodman and Newell 1985, Ellstrand 1988, National Academy of Sciences 1989). This realization raised the question, Will the flow of transgenes from crops into natural populations cause any problems? This issue has been raised by critics of crop biotechnology who are concerned about whether hybridization between transgenic crops and wild species and local land races will lead to the evolution of "superweeds," "genetic pollution," and the loss of valuable germplasm in centers of diversity (e.g., Van Aken 1999). In fact, the introgression of transgenes into natural populations is the most frequently discussed potential environmental impact of plant biotechnology (e.g., Ellstrand 2001, Hails 2000, Lutman 1999, Marvier 2001, Snow and Moran-Palma 1997, Wolfenbarger and Phifer 2000). Such concerns are heightened by the fact that pollutants such as noxious chemicals or radioactivity typically decay over time, while unwanted engineered genes do not necessarily decay in the environment. Transgenes may actually increase in frequency over time if the organisms in which they reside reproduce.

But the impact of crop genes in wild populations is an issue for any domesticated plant variety, transgenic or not. Therefore, historical experience with conventionally improved crops can provide valuable lessons for predicting the likelihood of escape of transgenes into the wild and the consequences of their presence in the wild. Understanding the environmental impacts of escaped transgenes requires first an understanding of the movement of domesticated genes into natural populations and their impact on those populations.

What are the specific circumstances under which domesticated alleles are able to enter natural populations? Generally, it is supposed that such spontaneous hybridization typically occurs when pollen from a domesticated plant fertilizes a wild plant, which then bears the hybrid seed or seeds (Ladizinsky 1985). The pollen (male) parent may be a plant in cultivation, a volunteer left from a previous planting, a recent escape from cultivation, or even a plant growing along the roadside that germinated from a seed that accidentally spilled off a truck transporting grain. We can suppose that pollen from this plant finds its way (generally borne by an insect or the wind, depending on its pollen dispersal syndrome) to the flower of a wild relative. The wild plant acting as the seed (female) parent may be growing next to the domesticated plant, in an adjacent field, or, as discussed in chapter 4, at a surprising distance from the pollen source.

An alternative hybridization pathway may involve a wild plant acting as a pollen parent and a cultivated plant as a seed parent. In this case pollen finds its way to the cultivated plant. The resulting hybrid seed may either drop to the ground and germinate or be intentionally planted. If the hybrid finds itself germinating and growing among or near other natural hybrids or genetically pure wild plants, it can now serve as a pollen parent, and genes from domesticated plants can still make their way into natural populations. Note that this latter scenario is the situation described for "The Case of the Bolting Beets" in chapter 1.

More complicated scenarios are possible. Plant breeders sometimes use a third species as an intermediate mate, a genetic "bridge" to move genes from one species to another. The same could occur in the field. One wild species could serve as a reproductive bridge to another. Suppose a certain crop might be easily able to cross with wild species A but not wild species B. However, species A and species B are easily able to cross with each other. Thus, natural hybridization between the crop and species A delivers crop genes into populations of species A. Individuals with a hybrid ancestry in those populations can mate with species B, delivering the crop genes to that species. And then, perhaps, those individuals can mate with species C. And so on.

It is even possible for natural hybridization to occur under "unnatural" conditions. For example, spontaneous hybridization between a crop plant and a wild species that were both deliberately planted in a botanical garden led to the creation of a new subspecies, *Solanum* × *edinense* ssp. *edinense* (Hawkes 1990; see chap. 7).

Here's the basic terminology for naming hybrids and immediate hybrid descendants. If mating between domesticated and wild plants is successful, the offspring are properly called "hybrids" or, in the geneticist's shorthand, "F_1s." Sometimes such progeny are called "first-generation hybrids" to distinguish them from later-generation individuals that are more remote from the initial hybridization event. If first-generation hybrids are fertile and have progeny of their own, these offspring have specific designations based on the identity of the parents. Hybrid-by-hybrid ($F_1 \times F_1$) matings (including F_1 self-fertilizations) yield "F_2" progeny (also called "segregants" or "second-generation hybrids"). But if a first-generation hybrid crosses with one of the parental species, the progeny are said to be "first-generation backcrosses" (designated "B_1s"). A mating between a first-generation backcross and the same parental species creates "second-generation backcrosses" (designated "B_2s"). Additional generations of backcrosses are named "B_3s," "B_4s," and so on. Individuals whose genetic history involves multiple backcrossing events over multiple generations are called "introgressed" or deemed to be "introgressants."

Geneticists who conduct controlled crosses tend to be careful about this terminology. But some field scientists who encounter plants in nature that appear to have some hybrid ancestry are frequently less precise, using the term "hybrid" loosely to describe plants that could be F_1s, F_2s, B_1s, or introgressants or resulting from any other genealogy that involves one or more hybridization events. Without genetic analysis it is impossible even to begin to identify their parentage. Plant systematists who recognize that populations of plants with apparent hybrid ancestry may contain individuals with different genealogies often describe those populations as "hybrid swarms" or "hybrid zones."

This book does not cover a related issue that has been the subject of many other publications. That issue is gene flow in the reverse direction, that is, the introgression of and subsequent incorporation of wild alleles into cultivated populations. Whether spontaneous wild-to-crop gene flow has played an important role in the evolution of crops grown under traditional agriculture has been the topic of considerable discussion and speculation for decades (e.g., Harlan 1965, Small 1984, Jarvis and Hodgkin 1999). Likewise, modern plant breeders have intentionally moved desirable genetically based traits from wild species into crops through deliberate hybridization and repeated backcrossing to the crop. Sometimes it is necessary for breeders to undertake "heroic" efforts to create hybrids, such as the rescue and culture of embryos that would otherwise be aborted (Sharma et al. 1996). Such techniques are largely laboratory or greenhouse exercises and are covered in modern crop improvement books (e.g., Fehr 1987).

Before discussing in detail what is known about spontaneous hybridization between domesticated plants and their wild relatives, the next few chapters lay out a background to provide a comparative context. For example, it is important to understand something about natural patterns of plant hybridization. The next chapter provides that overview, with a mini-review of natural hybridization in plants.

CHAPTER 3

Natural Hybridization between Plant Species

> Occasional hybridization between recognizable species . . . is the rule in flowering plants.
>
> —G. LEDYARD STEBBINS, PLANT EVOLUTIONIST, 1959

Didn't we learn that if populations hybridize, then they can't be species? Isn't hybridization too rare to make much of a difference? Does natural hybridization occur among species frequently enough to be interesting? This chapter examines the patterns of spontaneous hybridization between natural plant species to provide a context for understanding spontaneous hybridization between domesticated species and their wild relatives.

Hybridization Happens

Forget what you learned in Basic Biology class about species and hybridization. Simply stated, hybridization can occur between certain pairs of plant species. In fact, it makes sense in an evolutionary context. If species were fixed entities, we wouldn't expect hybridization to occur between them. But if species are constantly evolving, then the accumulation of barriers that prevent species from mating takes time. The time may be as brief as single generation but may be much longer, even hundreds of generations. Thus, at any given moment those barriers will be incomplete for certain organisms (Grant 1981).

In plants the situation varies tremendously. Whether and how much natural hybridization occurs varies with the particular pair of species involved and even the populations involved. If hybridization does occur, it might be restricted to producing a few hybrids. At the other extreme, continued intermating may homogenize the two species so thoroughly that they evolve to become a single evolutionary entity.

Clearly, not all pairs of plant species hybridize. Pine trees cannot mate with roses; apples cannot mate with oranges. While many pairs of closely related species are capable of hybridization, in many other cases closely related species are not capable of hybridization. Dozens of experimental studies have shown certain pairs of closely related species to be cross-incompatible; that is, the pollen of one species is incapable of fertilizing the other (e.g., Levin 1978, Grant 1981). The three species *Gilia transmontana, G. flavocinta,* and *G. malior* present one of the many examples. These small-flowered annual wildflowers of western North America are so closely related that they are difficult to distinguish from one another. However, experimental hand-crosses between them fail completely (Grant 1964a).

In fact, successful spontaneous hybridization depends on a variety of factors (Levin 1978). First, cross-pollination must occur. For this to happen, both parental species must be flowering at the same time. The plants must be close enough in space to allow a vector (such as wind, water, or an animal) to carry pollen from the male parent to the receptive tissues of the female parent. The pollen must be able to effect fertilization. Finally, the resulting embryos must develop into viable seeds and germinate. Thus, any one of several factors may prevent hybridization. Still, even if most pairwise combinations of plant species do not hybridize, natural hybridization need not be rare.

When two closely related plant species occur in a mixed population, it is not unusual to encounter a few plants that appear to be hybrids, with features that either combine those of the two co-occurring species or are intermediate to them. Consider, for example, the case of the oracle oak.

The California black oak, *Quercus kelloggii,* is a tree of the woodlands and forests of California's mountains. Its deeply lobed leaves have what those of us who grew up in the northeastern quadrant of the United States would call a typical oak-leaf shape. Those leaves are deciduous; in the autumn they turn from green to any color from yellow to rich brown, depending on the tree. The interior live oak, *Q. wislizenii* var. *frutescens,* is a shrub of warmer and drier microclimates, typically chaparral. This species' leaves are small, leathery,

Fig. 3.1 Leaves of the California black oak, *Quercus kelloggii (top),* interior live oak, *Q. wislizenii (bottom),* and their hybrid, the oracle oak, *Q.* × *morehus (center).* Photo by Janet Clegg. Used with permission.

spine-toothed or with a smooth edge, and evergreen (Hickman 1993), what native Californians would call the shape of a typical oak leaf. When the two species occur in mixed stands, oak trees of small stature with shallowly lobed, semideciduous leaves are occasionally present as well. These intermediate trees have been named the oracle oak, *Q.* × *morehus,* and have long been thought to be hybrids of *Q. kelloggii* and *Q. wislizenii* (Jepson 1909). (See fig. 3.1.)

Nason tested whether oracle oaks were indeed hybrids of the other species (Nason et al. 1992). He visited a population of intermixed California black oaks, interior live oaks, and a few oracle oaks in California's San Jacinto Mountains. He collected leaves from presumably "pure" *Q. kelloggii* and *Q. wislizenii* as well as from thirteen individuals *Q.* × *morehus.* He analyzed those leaves with starch gel electrophoresis to visualize the genetically controlled markers known as allozymes (also known as isozymes). As discussed in detail in chapter 5, genetically controlled markers are good tools for testing whether individuals are hybrids or have hybrid ancestry. Each plant gene (locus, plural "loci") has two copies (alleles), one from the mother and one from the father. If two species differ for alleles at a given locus, we would expect that a hybrid between two different species would have the two different alleles specific to the parental species. When two different alleles are present at a locus, the genotype of that locus is said to be heterozygous; if the two alleles are the same type, the locus is said to be homozygous.

After genetic screening of the pure black oaks and live oaks, Nason found, for each of the seven loci surveyed, at least one allele present in one species but not in the other. For all thirteen oracle oaks, "at least one locus bore an allele unique to *Q. kelloggii* [and] at least one locus bore an allele unique to *Q. wislizenii*" (Nason et al. 1992). That pattern is concordant with recent hybrid ancestry. In fact, all trees had genotypes that were indistinguishable from first-generation hybrids. In no case was a locus homozygous for an allele specific to one of the parental species. Analysis by a statistical model revealed that all the oracle oaks analyzed were in fact first-generation hybrids (Nason and Ellstrand 1993). Apparently the hybrids are unable to create their own progeny, preventing subsequent introgression.

The study of the oracle oaks is just one of hundreds documenting interspecific hybridization between otherwise discrete plant species. Well-studied cases of natural hybrids, that is, those that are supported by more evidence than simple morphological intermediacy, number in the hundreds (Arnold 1997, Ellstrand et al. 1996, Grant 1981, Stace 1975b). The great majority of these cases involve hybrids between close relatives, that is, members of the same genus. Additionally, in certain plant families (e.g., grasses, cacti, orchids) hybridization between members of different genera is not rare.

Nonetheless, even if hybridization is common, it is not ubiquitous. Ellstrand et al. (1996) evaluated the taxonomic distribution of natural hybridization in five different floras and found that, on average, about 11 percent of the species were involved in spontaneous hybridization. Furthermore, they found that the incidence of natural hybridization occurs unevenly over plant genera and families. In each region most of the reported hybrids occur within a handful of families. For example, in the Great Plains flora (Great Plains Flora Association 1986), five families (Asteraceae, Poaceae, Rosaceae, Fabaceae, and Amaranthaceae) contribute more than half of the total reported hybrids. But one of the most important families in that region, the Cyperaceae, represented by 113 species, did not contribute a single hybrid. Furthermore, most of the hybrids reported for a family typically occur in a single genus. For example, in the British Isles, forty-six different hybrids are reported for the Onagraceae, with forty-three of those occurring within *Epilobium* (Stace 1991).

The amount of species intermixing also varies among different pairs of hybridizing species. In the case discussed above, hybridization between *Q. kelloggii* and *Q. wislizenii* appears to be limited to the production of a few first-generation hybrids thinly scattered among mixed stands of the parents. We would expect this to be the case if F_1s are highly sterile. For plants, F_1s typically have some reduced fertility but are rarely fully sterile (Grant 1981, Stace 1975b). In fact, many plant hybrids do not suffer reduced fitness, and in certain cases they exhibit increased fitness relative to either or both of the parental taxa (e.g., Arnold 1997, Hauser et al. 1998b, Klinger and Ellstrand 1994).

Genetic analysis of natural hybrids suggests that the formation of first-generation hybrids without continued introgression is apparently the exception and continued introgression is the rule. Studies employing allozymes and DNA-based genetic markers have revealed dozens of instances of natural introgression in plants (Rieseberg and Wendel 1993). For example, a population of two manzanita species, *Arctostaphylos viscida* and *A. patula,* in the Yosemite region of California has received decades of attention from geneticists and ecologists (e.g., Ball et al. 1983, Dobzhansky 1953, Epling 1947) because the plants represent the full spectrum of phenotypes from one species to the other. Elsewhere the two species are readily distinguished by a number of reproductive and vegetative characters (Hickman 1993). Allozyme data were collected for fifty-six plants in the Yosemite population and compared with genotypes of nearby pure populations of parental species (Ellstrand et al. 1987). A statistical analysis assigned about 60 percent of the sample to one or the other of the pure species (Nason et al. 1992). If true F_1s were present, they were too rare to be detected; the remaining individuals appeared to be mostly backcrosses, especially backcrosses to *A. patula* (Nason et al. 1992). Because that statistical analysis cannot assign individuals to generations beyond the second generation of hybridization, those individuals assigned as backcrosses are not necessarily first-generation backcrosses or even true backcrosses. Nonetheless, it is clear that in the case of the Yosemite manzanitas, interspecies mixing has proceeded much further than in the case of the oracle oaks. Even though about 40 percent of the Yosemite manzanitas have hybrid ancestry, this example represents roughly a midpoint in the spectrum of how extensive interspecies hybridization can be.

In certain cases hybridization and introgression can be so extensive that parental phenotypes are exceedingly rare. For example, radish, *Raphanus sativus,* and jointed charlock, *R. raphanistrum,* behave as discrete species in Europe. *R. sativus* has white, pink, or purple flowers and a swollen taproot (Panetsos and Baker 1967). Its fruits (called "loments") are smooth, spongy, and corky, easily crushed by hand. Jointed charlock has white or yellow flowers, a taproot with frequent branching, and fruit with constrictions between the seeds that is "not crushable, except between the seeds, by hand pressure" (Panetsos and Baker 1967). In some European countries the two species sporadically hybridize when in proximity, leading to localized hybrid swarms (e.g., in the United Kingdom; Stace 1975b), but in other countries (e.g., in the Netherlands; deVries et al. 1992) they apparently do not hybridize. The situation in California is radically different. There, jointed charlock and radish have hybridized and introgressed into a single polymorphic coalescent complex extending over hundreds of square kilometers in California's coastal plains and inland valleys (Panetsos and Baker 1967). Nowadays, in most locations, it is impossible to find individuals that can be clearly assigned to *R. sativus* or *R. raphanistrum* (N. Ellstrand, J. Nason, J. Clegg, and S. Hegde, unpublished data). Instead, most individuals have a taproot characteristic of *R. raphanistrum* and fruits with some intermediacy between characters of the two *Raphanus* species, typically somewhat crushable but with some constrictions. Flower color is typically polymorphic in most populations. Pink, purple, yellow, and white flowers are generally all present and do not co-vary with fruit or root morphology. In this case, in populations comprising thousands of hectares, plants of hybrid ancestry have swamped their genetically pure progenitors to near extinction. Other examples of coalescent hybrid complexes covering large areas include Australia's introduced thistles, *Onopordum acanthium* and *O. illyricum* (O'Hanlon et al. 1999), and Germany's native violets, *Viola riviniana* and *V. reichenbachiana* (Neuffer et al. 1999).

Even hybridization between the same pair of species can vary from location to location. For example, plants of apparent hybrid ancestry between California's white sage, *Salvia apiana,* and black sage, *S. mellifera,* are often found when the two species are intermingled. Nonetheless, when Meyn and Emboden (1987) visited twenty sites containing both species, they found no evidence of hybridization at three of them. At nine of the remaining sites, hybridization was limited to apparent F_1s; at the other eight, the authors found individuals that were apparently both F_1s and introgressants.

The amount of introgression can vary among different genes as well. That is, for some loci the alleles from parental species A appear to have great difficulty persisting in the hybrid lineage past the first generation, while for other loci alleles from species A introgress easily into future generations, sometimes introgressing into otherwise genetically pure populations of species B. Therefore, although morphological intermediacy and molecular confirmation of introgression often go hand in hand, in other cases one or a few introgressed genetic markers may be found in otherwise morphologically pure individuals. Consider the following example.

The sunflower, *Helianthus petiolaris*, was introduced into southern California less than a century ago. As its range began to expand, it encountered populations of the previously established *H. annuus*. Hybridization occasionally occurs between the two species. Southern California *Helianthus* populations were analyzed for genetic markers specific to both species (Dorado et al. 1992). All but 4 of the 141 individuals of morphologically "pure" *H. petiolaris* sampled had the chloroplast genotype of *H. annuus,* but only two of those plants had ribosomal DNA markers characteristic of *H. annuus.* Interestingly, no introgression was observed in the other direction; that is, plants looking like pure *H. annuus* were not found to have any markers specific to *H. petiolaris*.

Even the geographic scale of introgression can be locus-specific. Where *Yucca schidigera* and *Y. baccata* come into contact in Arizona, plants of intermediate morphologies occur over a narrow hybrid zone less than a few dozen kilometers wide. Both species were analyzed for a number of genetic markers over their ranges. For one DNA-based marker (elucidated by the RAPD technique, described in chap. 5) the geographic pattern observed matched the same pattern observed for plant morphology, shifting from high frequency in apparently pure *Y. schidigera* to low frequency in apparently pure *Y. baccata* (Hanson 1993). But five other DNA-based markers and two chloroplast markers specific to *Y. schidigera* were decoupled from morphological traits, occurring over a much greater distance into the range of *Y. baccata* (in one case on the order of more than 150 kilometers). A similar situation has been reported for two species of cottonwood, *Populus fremontii* and *P. angustifolia,* in northern Utah (Martinsen et al. 2001). Dozens of cases of such differential introgression have now been detected with molecular markers (e.g., Rieseberg and Soltis 1991).

In conclusion, natural hybridization and introgression between plant species is idiosyncratic. They vary with the specific pair of species involved. They vary among populations. Hybridization products may be limited to first-generation hybrids or a local hybrid swarm. The introgression of individual alleles may be limited to first- or second-generation individuals while other alleles readily introgress from one species into another. Mixing of lineages may be so extensive that a mongrelized population replaces one or both parental types. And at the extreme, hybridization may result in the evolution of new species.

Hybrid Speciation in Plants

Hybridization appears to play an important role in plant speciation. More than 70 percent of plant species may have hybrid ancestry (Grant 1981). This often-quoted value is a bit misleading. Having hybrid ancestry is not the same as directly resulting from a hybrid event. That is, a single hybrid event may lead to the creation of a single species, and then that species may diversify into many more, all of which may have the genetic signature of a hybrid origin.

Nonetheless, dozens of cases of interspecific hybridization leading to speciation in plants have been supported with genetic evidence (e.g., Grant 1981). The list includes several important crops, including wheat, sweet potato, and groundnut (Smartt and Simmonds 1995). In addition to those that evolved thousands or millions of years ago, many cases of new plant hybrid species have been reported within recent history. Abbott (1992) listed ten plant taxa that evolved in the last few hundred years as a consequence of hybridization between an introduced exotic and either another exotic or a native species.

Perhaps the most notorious example is the salt marsh grass *Spartina anglica* in the British Isles (Gray et al. 1991, Thompson 1991). The native salt marsh grass of the British Isles is *S. maritima*. This species has thirty pairs of chromosomes (by convention, cytogeneticists report that number as "2n = 60"). The *S. alterniflora* of the New World was introduced into the British Isles about the mid-1800s. This species has a different chromosome number (2n = 62). The two species hybridized; the hybrid, *S. × townsendii* (2n = 61), was first identified in 1879. As in the case of many hybrids with mismatched chromosomes that could not pair normally during gamete formation, *S. × townsendii* was sterile. Despite inability to set seed, the hybrids, like their parents, were capable of vegetative reproduction by rhizomes and began to spread. By 1892 the seed-producing species *S. anglica* had evolved from the sterile hybrid by chromosome doubling (2n = 122 typical for this species). The appearance of this new species is more than an academic curiosity. *S. anglica* has become a spectacularly successful invasive species, now occupying "approximately 10,000 hectares along the coast of Britain" (Thompson 1991). Despite its recent evolution, this species has radically altered the ecology of Britain's coasts, invading the open intertidal flats, replacing more diverse native plant communities, altering succession, and limiting the food supply of birds that forage in those habitats.

The stabilization of hybridity associated with duplication of chromosomes is called "allopolyploidy" (also known as "amphiploidy"). A generalization of the evolution of allopolyploidy is illustrated in figure 3.2. If the hybrid species has the four genomes of two parental species, it is called an "allotetraploid" (in which case 2n = 4x); if it has six genomes of three parental species, it is called an "allohexaploid" (2n = 6×). Allopolyploidy is probably the most common form of hybrid speciation in plants (Grant 1981, Soltis and Soltis 1999).

But hybrid species can arise by means of other evolutionary tricks. For example, many hybrid lineages can persist as plants that produce seed without sex, a process known as "agamospermy" (Richards 1997). This situation is known for a wide variety of plants. A well-known case is the dandelion genus *Taraxacum*, which comprises hundreds of asexual species (all or most are thought to be hybrids) and only a handful that are sexual. A third example of hybrid speciation involves the evolution of sexual lineages that combine some genes of both parental species but are reproductively isolated from both parents, sometimes called "recombinant species" (Grant 1981). While more than fifty cases of recombinational speciation have been proposed, only about eight cases have been studied well enough to provide substantial proof of the phenomenon (Rieseberg 1997). The best-documented cases involve three sunflower species (Rieseberg 1991a) and a species of iris (Arnold 1993). Other mechanisms of hybrid speciation in plants include the evolution of complex cytogenetic miracles called "permanent odd polyploidy" and "permanent translocation heterozygosity" (the interested reader is referred to Grant 1981).

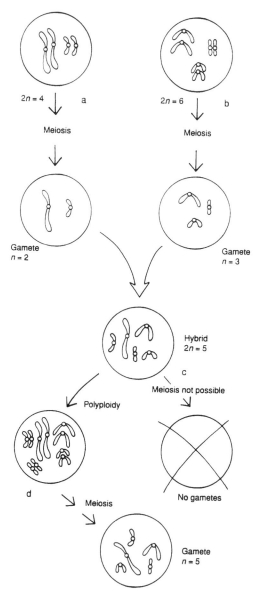

Fig. 3.2 Evolution of allopolyploidy. Species *a* and species *b* hybridize. Their hybrid *(c)* is pollen- and seed-sterile because its chromosomes cannot pair with one another during meiosis to create gametes. After spontaneous chromosomal doubling *(d)*, chromosomes are able to pair and create gametes. The polyploid individual has recovered sexual fertility. From fig. 1.31 of B. B. Simpson and M. C. Ogorzaly, *Economic Botany: Plants in Our World,* 3rd ed. © 2001 McGraw-Hill. Reproduced with permission of The McGraw-Hill Companies.

More information on natural hybridization in general, and in plants in particular, is available elsewhere. Arnold's (1997) book *Natural Hybridization and Evolution,* the appropriate chapters in Grant's (1981) *Plant Speciation,* and reviews by Rieseberg and colleagues over the last decade (e.g., Rieseberg 1995, 1997, Rieseberg and Carney 1998, Rieseberg and Ellstrand 1993, Rieseberg and Soltis 1991, Rieseberg and Wendel 1993) are all well written and informative.

CHAPTER 4

Evolutionary Consequences of Gene Flow—and Applied Implications

A little gene flow can go a long way.   —ANONYMOUS

The previous chapter makes it clear that natural hybridization is not unusual for plants. But is hybridization ever more than a biological novelty? When does it make a difference in the evolutionary fate of a population or a species? And will those changes ever have any implications for the plant populations that humans manage?

If domesticated plants mate with wild relatives, domesticated alleles have the opportunity to move into wild populations. What is the evolutionary impact of the arrival of new alleles via hybridization and introgression? The interplay of the four evolutionary forces-gene flow, natural selection, genetic drift, and mutation-determines whether and how evolution will proceed. Generally, gene flow and mutation add genetic variation to a population while selection and drift tend to deplete that variation. Population geneticists study how these forces interact and the evolutionary outcome of their interactions. This chapter uses the theoretical tools of population genetics to predict when the successful arrival of domesticated alleles in the wild makes an evolutionary impact. The predicted evolutionary impacts have logical applied implications that are discussed briefly. In later chapters we will see whether and how those predictions, both evolutionary and applied, are realized. Readers seeking a more thorough education in evolutionary genetics should consult a good evolution or population-genetics textbook (such as Futuyma 1998, Hartl and Clark 1997, or Hedrick 2000; see also the first five chapters of Hancock 1992).

The successful incorporation of genes from one population into another is the definition of gene flow (Futuyma 1998). In plants, gene flow (also known as "migration" or "immigration") occurs by successful pollination, seed dispersal, and even the dispersal of vegetative propagules (such as the runners of strawberry plants or the "joints" that fall off cacti). Because this book examines the consequences of mating between plants, the following discussion focuses primarily on gene flow by pollination.

Gene Flow

Gene flow can be a potent evolutionary force. A surprisingly small amount of gene flow can play a big role in evolutionary change (Slatkin 1987). The significance of gene flow generally depends on its magnitude relative to the other forces. The magnitude of gene flow among natural plant populations is idiosyncratic. It varies among species, populations, individuals, and even years (Ellstrand 1992). Gene flow by pollen may be effectively zero between plant populations that have strong physiological crossing barriers, create sterile progeny, are isolated by great distances, or do not overlap in flowering time. At the other extreme, gene flow may be so extensive that most seeds set in one locale have pollen parents that live in another (e.g., Dow and Ashley 1998, Friedman and Adams 1985, Nason et al. 1998). Even when sexually compatible species with similar flowering times grow near one another, levels of gene flow may vary with a number of ecological and genetic factors. Similarly, levels of gene flow from domesticated plants to their wild relatives are expected to be highly variable and to depend on a variety of factors.

The role of spatial distance is an important one. In the case of pollen, for the vast majority of plants studied, most pollen disperses very close to the source, and the dispersal curve has more events concentrated in its peak and has a longer tail than a normal curve. Statisticians characterize that kind of curve as "skewed and leptokurtic" (Levin and Kerster 1974; see also fig. 4.1). Seed dispersal in plants tends to follow the same shaped curve as that of pollen. Short-distance pollen dispersal tends to be predictable, dropping off rap-

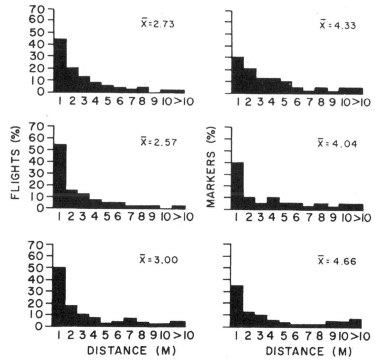

Fig. 4.1 Typical "skewed and leptokurtic" patterns of pollen dispersal, whether estimated by pollinator flight distances *(left)* or genetic markers from progeny testing
(right). These particular data were collected at different times in the same population
of the Texas wildflower *Phlox drummondii*. From fig. 1 of D. A. Levin, "Dispersal versus
Gene Flow in Plants," *Annals of the Missouri Botanical Garden* 68 (1981): 233–253. Reproduced with permission of the author.

idly at distances from a source on the order of one meter to a dozen meters.
Long-distance dispersal events on the order of fifty meters to thousands of meters can have considerable biological significance but are much less predictable and much more difficult to measure than short-distance events (Ellstrand
1992, Ouborg et al. 1999). Experimental studies have shown that intersite and
interplant variation in gene-flow rates becomes more pronounced at long distance (Klinger and Ellstrand 1999). Apparently, factors other than distance become much more important in determining gene-flow rate at this spatial
scale.

One such factor is the size of the gene-flow source populations relative to
the recipient sink population. Generally, if two populations, one large and

one small, are exchanging genes, it is expected and has been shown that gene flow will be asymmetric. The small population typically receives higher levels of gene flow than the large one (reviewed by Ellstrand and Elam 1993, Handel 1983, but see Klinger et al. 1992). For example, Ellstrand et al. (1989) used genetic markers to analyze the paternity of seeds set in three small experimental stands of wild radish (fifteen individuals each) planted a few hundred meters apart from one another. They did not detect gene exchange among the experimental populations, but they did detect gene flow into two of them from very large natural populations composed of hundreds of thousands of plants, even though those populations were at a distance of more than a thousand meters.

In the case of domesticated plants and their wild relatives, the relative numbers of each should play a big role in the direction and amount of gene flow. The cultivation of one or a few fruit trees in a backyard or dooryard garden close to where a wild relative grows in abundance probably results in very low levels of cultivated pollen movement into the wild population. However, under those circumstances the wild trees probably sire a substantial fraction of seeds set by the cultivated plants. In contrast, an agronomic crop like wheat, maize, or soybeans growing under the practices of modern industrial agriculture may contain millions of plants. Nearby populations of related weeds that are closely controlled with herbicides may contain only a handful of individuals. In such situations we might expect little gene flow into the crop but higher levels of gene flow into weed populations, even if they are hundreds or thousands of meters away.

Many experiments have now been performed to measure spontaneous hybridization rates between crops and cross-compatible wild relatives; representative examples are listed in table 4.1. Typically these experiments involve a stand of crop plants surrounded by synthetic populations of a wild relative. The plants are selected so that the crop bears an allele that is absent in the wild plants. Progeny testing of seed harvested from the wild plants reveals those that bear the allele specific to the domesticate. That fraction of progeny is the estimate of the hybridization rate. The variation in gene-flow rates measured in such studies (table 4.1) is as dramatic as that measured for natural populations.

A few experiments have shown certain wild-domesticate pairs to be apparently fully incompatible. Field experiments were conducted to test whether potato *(Solanum tuberosum)* spontaneously mates with the related weed species *S. nigrum* and *S. dulcamara*. McPartlan and Dale (1994) planted both spe-

Table 4.1 Some Experimental Estimates of Domesticated-to-Wild Gene-Flow Rates

Cultigen	Scientific Name	Wild Relative	Hybridiza-tion Rate	Range of Distances (m)	Maximum Distance (m) at Which Gene Flow Was Detected	Citation
Bread wheat	Triticum aestivum	Aegilops cylindrica	1–7%	Intermingled	n/a	Guadagnuolo et al. 2001a
Cocona	Solanum sessiliflorum	Same sp.	0.43–100%	1–9.0	9	Salick 1992
Foxtail millet	Setaria italica	S. verticilliata	0.50%	0.4	0.4	Till-Boutraud et al. 1992
Foxtail millet	Setaria italica	S. viridis	0.15–0.27%	0.17–0.4	0.4	Till-Boutraud et al. 1992
Pearl millet	Pennisetum glaucum	P. sieberanum	39%	Intermingled	n/a	Renno et al. 1997
Pearl millet	Pennisetum glaucum	P. violaceum	8%	Intermingled	n/a	Renno et al. 1997
Potato	Solanum stenotomum	S. sparsipilum	95%	Intermingled	n/a	Rabinowitz et al. 1990
Potato	Solanum tuberosum	S. dulcamara	0%	0–20	n/a	McPartlan and Dale 1994
Potato	Solanum tuberosum	S. nigrum	0%	0–20	n/a	McPartlan and Dale 1994
Radish	Raphanus sativus	Same sp.	0–100%	1–1,000	1,000	Klinger et al. 1991
Radish	Raphanus sativus	Same sp.	0–79%	1–400	400	Klinger et al. 1992
Rapeseed	Brassica napus	B. campestris	9–93%	Intermingled	n/a	Jørgensen et al. 1996
Rapeseed	Brassica napus	Sinapis arvensis	0%	Intermingled	n/a	Lefol et al. 1996a
Rice	Oryza sativa	Same sp.	1–52%	Intermingled	n/a	Langevin et al. 1990
Sorghum	Sorghum bicolor	S. halepense	0–100%	0.5–100	100	Arriola and Ellstrand 1996
Squash	Cucurbia pepo	C. texana	5%	1,300	1,300	Kirkpatrick and Wilson 1988
Sugar beet	Beta vulgaris	Same sp.	1%	0–210	210	Vigouroux et al. 1999
Sunflower	Helianthus annuus	Same sp.	0–27%	3–1,000	1,000	Arias and Rieseberg 1994
Sunflower	Helianthus annuus	Same sp.	4–42%	3–400	400	Whitton et al. 1997

cies around a plot of potatoes transgenic for resistance to the antibiotic kanamycin. Seeds were harvested from the weeds and germinated. Thousands of seedlings were screened; not one showed resistance to kanamycin. The measured hybridization rate was zero.

However, reports of experimentally measured crop-to-wild hybridization rates of zero are exceptional. In many cases hybridization rates exceed 1 percent, for cultivated-wild pairs as diverse as domesticated sunflower and wild sunflower (Arias and Rieseberg 1994), sorghum and johnsongrass (Arriola and Ellstrand 1996), squash and wild gourd (Kirkpatrick and Wilson 1988), domesticated radish and wild radish (Klinger et al. 1991), and various cultivars of rice and red rice (Langevin et al. 1990). Studies measuring how gene flow changes with distance (e.g., Arias and Rieseberg 1994, Klinger et al. 1991, Salick 1992) generally find the skewed and leptokurtic dispersal curves that are typical for plants in general.

And, as expected with leptokurtosis, the dispersal tail can be very long indeed. Hybridization rates greater than 1 percent have sometimes been detected over distances of hundreds of meters (e.g., Arias and Rieseberg 1994, Kirkpatrick and Wilson 1988, Klinger et al. 1991). Indeed, for many of the studies listed in table 4.1 the real maximum dispersal distance might have been underestimated. One reason is that the maximum dispersal distance identified was often the same as the maximum distance sampled (see Ellstrand 1992 on how this experimental design underestimates gene flow by truncating the potential dispersal curve). Furthermore, it simply becomes harder to measure gene flow emanating from a source at great distances because it becomes increasingly hard to sample the increasing perimeter adequately. Consider placing one hundred plants as potential pollen recipients, each one meter wide, at one hundred meters from a point source of pollen. Those plants would sample about $100/(2\pi100) = 1/(2\pi) =$ roughly 16 percent of the perimeter. At one thousand meters from a point source, the same number of plants would sample only 1.6 percent of the perimeter. If gene flow is relatively rare at this distance, it might easily be missed.

At shorter distances crop-to-wild hybridization rates may exceed 50 percent (e.g., Arias and Rieseberg 1994, Klinger et al. 1991, Langevin et al. 1990, Rabinowitz et al. 1990). But in other cases hybridization rates have been found to be quite low, even at short distances. Till-Bottraud et al. (1992) measured bilateral hybridization rates between foxtail millet, *Setaria italica,* and selected wild *Setaria* species growing in an experimental garden. At distances that were mostly less than one hundred centimeters they sometimes detected hybridization, but at rates that never exceeded 1 percent (from 0 to 0.58 percent).

Likewise, crop-to-wild gene-flow rates may vary tremendously among plants and plots. In an experiment involving a range of distances and target population sizes, gene flow from cultivated radish to wild radish varied from 0 to 79 percent, depending on treatment (Klinger et al. 1992). The results from similar experiments are summarized in table 4.1. As noted above, gene-flow rates tend to become particularly variable at great distances. That variability will tend to result in the underestimation of gene flow by averaging gene flow over plants sampled because most sampling schemes will be too small to detect those relatively rare plants that produce a large amount of seed sired by a distant mate (Klinger 2002).

Even if they often underestimate gene flow, these hybridization rates present a context for evaluating when gene flow by hybridization will make a difference in molding the genetic structure of a population. In the absence of the other three evolutionary forces, the primary evolutionary consequence of gene flow is to genetically homogenize populations (reviewed in Slatkin 1987). Thus, with generations of continued gene flow from an adjacent domesticated relative, wild populations should become genetically more croplike. It is generally held that individual crop cultivars typically contain substantially less genetic variation than individual populations of their wild relatives (Ladizinsky 1985, Doebley 1989). When that is the case, then the evolutionary result of continued substantial gene flow from a single cultivar into a wild population should be a decrease in genetic diversity. However, in the short term, or under other circumstances, genetic diversity in the wild might well increase, as discussed below.

When will the other evolutionary forces prevent genetic homogenization by gene flow? The conditions for homogenization by gene flow depend on whether immigrant alleles are neutral (gene flow versus genetic drift), detrimental (gene flow opposed by natural selection), or beneficial (gene flow in concert with natural selection) in the context of the ecological and genomic environment of the population receiving those alleles (Wright 1969). The last phrase cannot be overemphasized; the adaptive value of a trait may vary tremendously with its environment. A genetically based trait that is crucial for survival and reproduction of a crop (or a weed) in a managed agroecosystem might prove fatal in a nearby unmanaged ecosystem. Also, gene flow might be opposed by mutation (gene flow versus mutation).

Table 4.2 The Relative Importance of Recurrent Gene Flow versus Other Evolutionary Forces: Rules of Thumb

Gene Flow versus . . .	General Consequence of Gene Flow	Possible Significant Evolutionary Consequences*
Drift (neutral gene flow)	A few gene-flow events per generation are sufficient to eventually homogenize populations	Extinction
Negative selection (detrimental gene flow)	Gene flow counteracts selection of equal magnitude, resulting in local persistence of detrimental alleles	Extinction Reinforcement of reproductive isolation
Positive selection (beneficial gene flow)	Gene flow augments selection by accelerating the increase in frequency of a favored allele	Extinction Increased niche Evolution of increased weediness Evolution of invasiveness
Mutation	Unless gene-flow rate is less than or equal to the mutation rate, gene flow should overwhelm mutation effects	

*Under certain circumstances; see text.

In the next five sections I'll first pair gene flow with each of the other forces to examine the theoretical population-genetic consequences of their interactions. Then I'll briefly examine a selection of more complex interactions. The general expectations described in these sections are summarized in tables 4.2 and 4.3.

Consequences of Neutral Gene Flow

The Theory

When gene flow is absent and mutation rates are low, the evolution of neutral alleles (those not under natural selection) in a population depends on random processes (genetic drift). In the absence of other evolutionary forces, genetic drift, over time, will lead to genetic differentiation among populations, especially when effective population sizes are small (Wright 1969). "Small" is generally defined as fewer than one hundred to a few hundred individuals. "In

Table 4.3 General Consequences of Gene Flow from a Domesticated Plant into the Population of a Wild Relative

Adaptive Value of Novel Allele	Change in Genetic Diversity Immediately after One Bout of Gene Flow	Wild Population Genetic Structure Many Generations after . . .	
		One Bout of Gene Fow	Gene Flow Every Generation
Neutral	Increase	Novel allele persists at initial frequency	Genetic structure close to that of crop population
Detrimental	Increase	Novel allele extinct (i.e., no change)	Novel allele maintained in polymorphism
Beneficial	Increase	Novel allele replaces local allele(s)	Novel allele replaces local allele(s)

Note: Assumptions:
  Unilateral gene flow from a single crop population into a wild population
  The source crop population contains one or more alleles that are absent ("novel") in the recipient population
  Gene-flow rate is much less than 100 percent but much higher than the mutation rate
  Recipient population is too large for genetic drift to influence allele frequency change over time

a group of completely isolated populations, genetic drift alone would tend to fix different alleles in different local populations" (Slatkin 1987).

A little gene flow effectively nullifies the effects of genetic drift. The "rule of thumb" is that the arrival of about one immigrant every other generation, or one interpopulation mating per generation, should be sufficient to maintain the polymorphism of different neutral alleles at the same locus (Slatkin 1987). Interestingly, this relationship is independent of the size of the recipient population. Conservation geneticists often conclude that one migrant per generation (or a little more) will be sufficient to counteract the effects of drift, preventing strong divergence between populations and buffering against the loss of within-population variation (e.g., Allendorf 1983, Mills and Allendorf 1996).

What does this mean for the movement, establishment, and persistence of neutral alleles from domesticated plants into natural populations? A modest amount of gene flow—a few successful pollinations every generation that lead to adult plants—should readily introduce and maintain neutral alleles from domesticated plants in natural populations. This can occur even if the fraction of hybrid seed growing to reproductive maturity is vanishingly small compared with the total number of seeds set by the population. Because it is not unusual for one plant to create hundreds of seeds per year, a few successful interpopulation pollinations per generation in an entire plant population would often be impossible to detect without heroic efforts.

Evolutionary Consequences and Applied Implications

Fitness correlates of neutral gene flow are straightforward. If immigrant alleles were neutral in the wild population, then by definition we would expect no change in plant fitness after their establishment in that population.

Homogenization by neutral gene flow may or may not enhance local genetic diversity. Changes in local genetic variation depend (1) on whether gene flow is unilateral or bilateral, (2) on the genetic composition of the source population (in our case, the crop) compared with that of the sink population, and (3) on whether gene flow occurs only a few times or repeatedly over time.

Under traditional cropping systems and with close populations of wild relatives, gene flow may often be bilateral (Jarvis and Hodgkin 1999). In that situation neutral alleles might be expected to flow back and forth freely between wild and cultivated plants. If that is the case, allele frequencies in each population should converge. Changes in the diversity of the wild population would depend mainly on whether the crop and wild populations were initially genetically quite different from each other—in which case diversity should increase with the influx of novel alleles. If the two started with shared alleles, diversity would change only if the crop population was substantially more or less variable than the wild population. As noted above, the general view is that crop plants typically have relatively narrow genetic variability compared with their wild progenitors (Ladizinsky 1985).

Under modern industrial agriculture, gene flow should be unilateral because the large plantations under cultivation are not typically replaced by locally produced seeds but by seeds or plants created far from crop production areas. Thus, populations of wild relatives growing in the production area receive a constant stream of pure domesticated gene flow year after year. Crop cultivars used in modern industrial agriculture typically contain even less genetic variation than the same crops grown in traditional systems (Doebley 1989, Duvick 1984). Neutral gene flow from cultivars that hold very little genetic diversity over many generations should result, ultimately, in the reduction of diversity. If the crop is well differentiated from the wild relative, there will be a transient increase in diversity as novel crop alleles are introduced into the wild. Likewise, a single pulse of gene flow from a crop into a wild population will result in increased genetic diversity of neutral alleles. Unless genetic drift erodes that diversity, the increase should be permanent. If cultivars are replaced from time to time as new ones become available, it is possible that neutral allele diversity in the wild population may continue to accumulate, becoming an "archive" for "heirloom" cultivar alleles.

Under gene-flow pressure and in the absence of selection, a wild population would, over time, evolve to become increasingly croplike. In the extreme case, if all alleles at all loci that differentiate the domesticate from the wild population are effectively neutral and gene-flow pressure continues for enough time, the wild population would lose its genetic distinctiveness and become *identical* to the domesticated type. While such genetic swamping is usually assumed to occur under neutral gene flow (Rhymer and Simberloff 1996), it can also occur under beneficial gene flow (see below). Extinction by genetic swamping through repeated hybridization (sometimes called "extinction by assimilation") has recently caught the attention of conservation geneticists (e.g., Ellstrand and Elam 1993, Hopper 1995, Huxel 1999, Levin et al. 1996, Levin 2001, Rhymer and Simberloff 1996, Wolf et al. 2001). The situation is well known for animals, especially vertebrates.

The mallard duck provides a spectacular example (reviewed by Cade 1983, Rhymer and Simberloff 1996). The natural distribution of the mallard in the Northern Hemisphere has been expanding since the last ice age. Furthermore, humans have intentionally introduced the duck into new areas and unintentionally helped expand its range through habitat disturbance and modification. As its range expands, it encounters close relatives, sometimes hybridizing with them, apparently with no change in fitness. In certain cases the hybridization and introgression have been extensive. Almost all of the American populations of the once-distinct Mexican duck are now composed of mallard × Mexican hybrids and hybrid derivatives. The Mexican duck is essentially extinct. Hybridization with the mallard is a conservation threat to the black duck and mottled duck in the continental United States, the Hawaiian duck in Hawaii, and New Zealand's grey duck. Extinction by assimilation in animals is generally discussed as if it were the result of neutral gene flow rather than of the sweep of beneficial alleles. However, I am not aware of any study in animals that actually determines whether genetic swamping resulted from neutral, rather than beneficial, gene flow by comparing the fitness of the hybrids or hybrid derivatives with that of the genetically pure progenitor.

Hybridization as a contributor to increased risk of extinction is now known for many natural plant species (reviewed by Levin et al. 1996, Levin 2001). A notable example is the case of *Argyranthemum coronopifolium,* a rare endemic of the island of Tenerife (Levin 2001). Three of its seven known populations are in various stages of hybridization with the weedy and abundant *A. fructescens,* which expanded its range during the last century under human disturbance. One of the hybridizing populations was monitored for over thirty years; by 1996 *A. coronopifolium* was gone, replaced by hybrids, hybrid derivatives, and *A. fructescens.* As in animals, extinction by hybridization in plants is considered to result from the swamping effects of neutral gene flow (e.g., Stace 1975b). But fitness studies of extinction by hybridization in plants are rare. I am not aware of any study in plants that documents extinction by hybridization due to neutral gene flow. That may be because there is only a small window of time in which to study the process.

Extinction by assimilation can proceed with surprising rapidity. Even without any detrimental gene flow, under certain conditions the extinction of a population can occur after a handful of generations of hybridization with another species (Huxel 1999, Wolf et al. 2001). Wolf et al. (2001) developed a simulation model to examine the fate of hybridizing populations. They applied that model to a number of case studies, including hybridization between cultivated and wild sunflower (both *Helianthus annuus*), using realistic parameters. Their results under neutral gene flow are striking (note that they use the word "hybrids" to signify both hybrids and hybrid derivatives): "When the cultivar was sown for only a single generation, the fate of wild *H. annuus* depended on the initial frequencies of the wild and cultivated plants. If the initial population had equal numbers of each parental type ($N_1 = N_2 = 100$), the wild type persisted and hybrids were eliminated within a few generations (mean ± standard deviation = 4.1 ± 1.1). . . . But when the wild population was much smaller than the number of cultivated plants ($N_1 = 100$, $N_2 = 10,000$), as is likely to be the case in nature, the wild plants were completely replaced by hybrids 75% of the time, with a mean time to extinction of 4.0 ± 2.1 generations" (Wolf et al. 2001). In simulations where the cultivar was grown every year, with equal population sizes and equal competitive abilities, "hybrids replaced wild plants in 18.1 ± 4.5 generations" (Wolf et al. 2001). Thus, continued neutral gene-flow pressure from a domesticated species into a natural population may result in the eventual loss of genetic integrity of that population as it evolves into a genetic "shadow" of the domesticate.

Consequences of Detrimental Gene Flow

The Theory

It is often assumed that alleles that have been selected for during the domestication process are disadvantageous in the wild. For example, the closest wild relatives of maize bear their seeds singly, instead of attached to a solid cob. The cob is thought to prevent seed dispersal and presumably explains why free-living populations of maize do not occur (Martínez-Soriano and Leal-Klevezas 2000). Although domesticated alleles are commonly assumed to be maladapted in the wild, little if any empirical research has been conducted on the adaptive value of a specific crop allele relative to the corresponding wild-type allele in either managed or unmanaged environments. Also, keep in mind that whether an allele is detrimental, beneficial, or neutral may vary over time and space. For example, the fitness effects of an allele that confers pest resistance are apt to depend on whether the pest is present (e.g., Snow et al. 1998).

Population-genetic theory predicts that immigration of disadvantageous alleles can counteract local directional selection, reducing local fitness (e.g., Antonovics 1976, Lenormand 2002). Generally, detrimental alleles will be maintained in populations when $m \geq s$, where m is the fraction of the population replaced per generation by immigrants and s is the relative disadvantage of the immigrant allele (also known as the local selective coefficient) (Slatkin 1987). That is, moderate rates of incoming gene flow (ca. 1–5 percent per generation) are expected to be sufficient to introduce and maintain locally deleterious alleles when their relative disadvantage locally is of the same magnitude (i.e., 1–5 percent). If s is very small, that is, $\ll 1$ percent, for our purposes, the neutral expectations discussed in the previous section are more appropriate.

To calculate m from hybridization rates, the values must be halved to account for the fact that the paternal contribution makes up only half of an offspring's genotype. Even given that correction, it is clear that the typical crop-to-wild hybridization rates measured (table 4.1) are sufficient, in many cases, to counteract the effects of moderate local selection. That is, moderately harmful alleles will be maintained at some intermediate frequency in the population, simultaneously purged by selection and replenished by immigration.

Data from natural species support the expectation that only very strong local selection can counterbalance gene flow. Reciprocal transplant studies often reveal local adaptive differentiation in plant populations that are distant from one another (reviewed by Levin 1984, Waser 1993), but much less frequently at the microgeographic level at which substantial gene flow occurs (e.g., Antonovics et al. 1988, Waser and Price 1985). However, such differentiation becomes apparent if selection is very strong ($s > 0.3$–0.99; e.g., Antonovics and Bradshaw 1970, Linhart and Grant 1996, Lönn et al. 1996).

Evolutionary Consequences and Applied Implications

Clearly, the consequences of immigration of deleterious crop alleles in terms of altered local fitness and genetic diversity in wild populations depend on the relationship of m and s and the pattern of gene flow over time. If migration rates are about the same magnitude or greater than the selective disadvantage of the novel deleterious crop allele ($m \geq s$), natural selection will be unable to purge the disadvantageous allele as it is replenished by gene flow. In this situation, with continuous gene-flow pressure, both the original, well-adapted allele and the new, maladapted allele will coexist in the natural populations. Thus, with the immigration of a deleterious allele, mean population fitness should drop and genetic diversity should increase. A single pulse of gene flow would be expected to result in a transient decrease in fitness and a transient increase in diversity before selection brings the population back into equilibrium. If the relative disadvantage of the immigrant allele is much larger than the immigration rate ($s \gg m$), then the deleterious allele will not be able to establish in the natural population. In the latter case, fitness and diversity should change very little under either sporadic or continuous gene flow.

Adaptive differences between populations (in our case, adaptation to growth and reproduction under cultivation versus adaptation to growth and reproduction in unmanaged ecosystems) may lead to outbreeding depression, "a fitness reduction following hybridization" between populations (Templeton 1986). The fitness drop may be due to the fitness effects of the alleles themselves or due to deleterious *combinations* of alleles. One way this can occur is when hybrids are adapted to an environment that is intermediate between both well-adapted parents, and that environment does not exist. Alternatively, outbreeding depression may simply result from the evolutionary accumulation of genomic differences between lineages through isolation, leading to genetic disharmony in hybrids. In the latter case outbreeding depression is independent of adaptation to a given environment (Lynch and Walsh 1998). The classic sterility of the mule, an F_1 resulting from a horse-donkey cross, illustrates this second case.

But outbreeding depression can manifest itself much earlier in the life of a hybrid. For example, one-half or more of the hybrid seeds created by crosses between species of *Gilia* subsection Arachnion are aborted; in contrast, within-species crosses typically yield no aborted seeds (Grant 1964b, Grant and Grant 1960). In other words, outbreeding depression in hybrids may appear as a drop in the seed production of the maternal parent. The overall impact of outbreeding depression depends on its lifetime effects multiplied over all life-cycle stages from seed set of the hybridizing parent to viable seed and pollen production of the hybrid itself (e.g., Montalvo and Ellstrand 2001).

Outbreeding depression has been detected in many natural plant populations (Waser 1993), and naturally occurring interspecific hybrids often are partially or completely sterile (e.g., Stace 1975b). Sometimes, however, F_1s show a fitness boost, and outbreeding depression is delayed until as late as the F_3 generation ("advanced-generation breakdown"; e.g., Montalvo and Ellstrand 2001, Fenster and Galloway 2000).

Similarly, first-generation hybrids between domesticated plants and their wild relatives may have reduced fitness. For example, attempts by plant breeders to introduce commercially desirable alleles into crops from distant relatives (the secondary gene pool) are often constrained by the reduced fertility and viability of hybrids resulting from wide crosses (e.g., Simmonds 1981). Indeed, breeders have seen delayed outbreeding depression of the kind observed in some natural systems (e.g., in the F_2s resulting from hybridization among rice species; Oka 1957).

High gene-flow rates that result in aborted hybrid seeds, hybrids with very low survivorship, hybrid progeny with very low fertility, and other profound negative cumulative fitness effects over a life history may have dramatic effects on a population. Under this scenario the population receiving the gene flow may be unable to replace itself. Just like extinction by swamping discussed above, extinction by outbreeding depression through repeated hybridization has recently received the attention of conservation geneticists (e.g.,

Ellstrand and Elam 1993, Hopper 1995, Huxel 1999, Levin et al. 1996, Levin 2001, Muir and Howard 2001, Ribeiro and Spielman 1986, Wolf et al. 2001). The situation is best known for animals. For example, a Tatra mountain ibex *(Capra ibex ibex)* population went extinct in a single generation after interbreeding with ibexes of other subspecies that had been introduced with the goal of restoring the population size (Templeton 1986).

Extinction by outbreeding depression is probably important in plants as well. Many rare species of the British Isles hybridize with more common species (Stace 1991). A good fraction of those hybrids have reduced fitness (Stace 1975b). For example, the natural hybrid of the common horsetail *Equisetum hyemale* and the protected *E. ramosissimum* produces cones, but its spores are completely abortive (Duckett and Page 1975). But, to my knowledge, the role of depressed fitness in extinction by hybridization in a plant species has not yet been measured.

One reason for that research gap is that there may be little time to study the process. Extinction by outbreeding depression can proceed as rapidly as extinction by swamping. Theoretical models have demonstrated that, given the appropriate circumstances, the extinction of a population can occur after a handful of generations of hybridization with another species (Huxel 1999, Ribeiro and Spielman 1986, Wolf et al. 2001).

Evolution can be an alternative to extinction. Under gene-flow pressure that delivers unfavorable alleles, selection pressure to avoid hybridization is intense. Given the appropriate preexisting genetic variation and other favorable circumstances, a population suffering reduced fitness from hybridization may evolve strengthened reproductive isolation barriers. Such evolution of increased reproductive isolation in response to selection to avoid hybridization is sometimes referred to as "reinforcement" or the "Wallace effect" (cf. Grant 1981, Noor 1999). For example, certain grasses adapted to grow on heavy metal soils have evolved increased self-fertility and/or earlier flowering season compared with other plants of the same species (Antonovics 1968, McNeilly and Antonovics 1968). As a result, gene flow by pollen bearing deleterious alleles (intolerance to heavy metals) from progenitors growing just meters away is substantially reduced. Under the appropriate circumstances, evolution of reinforced isolation barriers can be rapid (Barton 2000), occurring in a handful of generations. While the evolution of reinforcement of isolation barriers is a potential response to detrimental gene flow, it is controversial, and the number of known cases remains small (Futuyma 1998, Noor 1999, Barton 2000, Turelli et al. 2001).

Consequences of Beneficial Gene Flow

The Theory

Certain alleles bred into crops may prove advantageous in the wild. Consider the following scenario. Both crops and their wild relatives suffer from the same pest species. Breeders have introduced a pest-resistance allele into the crop, but the wild plants have no genetically based resistance to those pests. All other things being equal, the pest-resistance allele from the crop would be expected to confer a fitness advantage to the wild plant. It is commonly assumed that traits like pest resistance and stress tolerance would be beneficial to plants in the wild (e.g., Ellstrand and Hoffman 1990).

However, little empirical research has been conducted regarding the adaptive value of a specific, potentially beneficial, crop allele relative to the corresponding wild allele in either managed or unmanaged environments. The area is a fertile one for research. A recent workshop on the ecological effects of pest-resistance genes in managed ecosystems could not identify a single experimental study directly examining whether pest-resistant alleles from a crop actually confer a fitness benefit in wild compared with native genotypes (Traynor and Westwood 1999).

Nonetheless, one recent study suggests that crop alleles conferring pest resistance can confer an advantage in the wild. Snow et al. (1998) compared the fitness of wild-crop sunflower hybrids with that of genetically pure wild individuals under field conditions. At their Ohio field site, 53 percent of the wild plants from North Dakota showed symptoms of rust, while none of the wild-crop hybrids showed symptoms. Plants not infected produced 24 percent more flower heads than infected plants. Apparently, one or more alleles from the crop confer some resistance to this disease.

When gene flow and selection work in concert, gene flow accelerates the increasing frequency of a favorable allele in a population. Wright (1969) modeled several situations involving gene flow in combination with favorable selection. His "Continent-Island" model, with one-way immigration from a large source population into a sink population, is one that will often be applicable to gene flow from crops to their wild relatives under the conditions of modern industrial agriculture. With modern cultivars, domesticated alleles may often be initially more or less fixed in the crop and absent from the wild population (especially if the crop was developed in a different region from the one in question). With these assumptions, Wright's model can be simplified to predict that the fraction of crop-to-wild immigrants, m, has the same effect on the speed of increase of the favorable alleles as the local selective coefficient, s (in this case s represents the relative *advantage* of the immigrant allele). That is, if the magnitude of both migration and the local selective coefficient are equal ($m = s$), then the rate of allele frequency change per generation is doubled relative to the situation with selection alone. If m is twice as great as s ($m = 2s$), then the rate will be tripled compared with the situation without gene flow. If s is very small, that is, < 1 percent, for our purposes, the neutral expectations discussed earlier in this chapter are more appropriate. Recall that to calculate m from a hybridization rate, it is necessary to divide it by two to account for the fact that immigrant pollen contributes only half of the genes residing in a hybrid. Even with that correction, it is clear that the crop-to-wild hybridization rates that have been measured (table 4.1) are often sufficient to significantly accelerate the incorporation of beneficial alleles into a population.

Under traditional agriculture and other conditions where plantations of domesticated species and populations of wild relatives are small and patchy, gene flow may be multilateral. Other models (cf. Wright 1969, Manasse and Kareiva 1991) may be more appropriate in these situations. The qualitative conclusions, however, are the same as for the Continent-Island model. The spread of a favored allele is enhanced by gene flow, and as gene flow increases, so does the speed at which the favored allele spreads.

Evolutionary Consequences and Applied Implications

The introduction of a new advantageous allele may result in a transient increase in diversity but eventually should result in the replacement of the inferior allele or alleles at that locus (e.g., Fisher 1937, Frank and Slatkin 1992). What is most important is that the immigration rate not be so low that drift and mutation pressure are able to counteract selection. Otherwise, it doesn't matter whether gene flow is sporadic or continuous, unilateral or bilateral; once a beneficial allele enters a population, it should spread to fixation. Therefore, the ultimate impact of the migration of a beneficial allele is no change in genetic diversity if the new allele replaces one preexisting allele or reduced genetic diversity if the new allele replaces multiple alleles.

Likewise, the migration of beneficial alleles also increases the mean fitness of the recipient population relative to other populations. Indeed, from a metapopulation perspective, gene flow will then accelerate the spread of that beneficial allele to other populations.

In the extreme case, if all alleles at all loci that differentiate the domesticate from the wild population are effectively beneficial, and gene flow is sufficient, the wild population would lose its genetic distinctiveness and evolve into the domesticated type. Such extinction by assimilation is usually assumed to occur under neutral gene flow (Rhymer and Simberloff 1996) but should occur more readily under beneficial gene flow (Huxel 1999, Wolf et al. 2001). As noted above, extinction by hybridization has recently caught the attention of conservation geneticists (e.g., Ellstrand and Elam 1993, Hopper 1995, Huxel 1999, Levin et al. 1996, Levin 2001, Rhymer and Simberloff 1996, Wolf et al. 2001) but is rarely discussed in terms of the immigration of beneficial alleles.

This is puzzling. Heterosis, a fitness boost after hybridization (also known as "hybrid vigor"), is well known for crosses within species. And recently it has been recognized that heterosis, while not the rule, is not rare in interspecific hybrids (e.g., Arnold 1997). One example involves natural hybrids between the invasive alien *Carpobrotus edulis* and the native *C. chilense* of coastal California. Vilà and D'Antonio (1998) compared the survival and growth of the two parental species and hybrids. The hybrids had a higher growth rate than both parents and a higher survival rate than *C. chilense*. Another example involves the invasive alien *Spartina alterniflora,* recently introduced into San Francisco Bay where it hybridizes with the native *S. foliosa*. In this case the hybrid has a higher male fitness than *S. foliosa,* with pollen that is more vigorous and abundant (Anttila et al. 2000). In both cases it is predicted that the hybrids are a threat to the native species—in terms both of competitively displacing the native and of mating with it, thereby hybridizing it to extinction.

And extinction by hybridization under beneficial gene flow can proceed very rapidly. Just as with neutral and detrimental gene flow, given the right conditions, the extinction of a population can occur in a handful of generations of hybridization with another species (Huxel 1999, Wolf et al. 2001). Thus, beneficial gene flow from a domesticated species into a natural population has the potential to convert that population into the domesticate.

In contrast to detrimental gene flow, which has the potential to thwart extinction through selection for increased reproduction isolation, beneficial gene flow should foster selection for decreased isolation. Those genotypes that allow for relatively high mating with plants with beneficial alleles will be favored. Thus, the evolution of the breakdown of reproductive isolation under beneficial gene flow will hasten extinction by swamping under hybridization.

Beneficial gene flow has the potential to lead to extinction, but it also has the potential to lead to adaptive evolution. New advantageous alleles can provide the opportunity for a species to expand its niche. The change can be subtle, or it can be dramatic, resulting in a new weed problem or a new invasive in natural ecosystems. The topic of hybridization as a stimulus for adaptive evolution is one that has received considerable discussion, especially for plants (e.g., Anderson and Stebbins 1954, de Wet and Harlan 1975, Lewontin and Birch 1966, Stebbins 1959, 1969, 1974). But empirical data are few, especially for plants. The cordgrass of the tidal flats of the British Isles, *Spartina anglica,* discussed in detail in the previous chapter, is perhaps the best-known case. This species is a hybrid derivative of a North American grass introduced to the United Kingdom *(S. alterniflora)* and a native species *(S. maritima)* (Gray et al. 1991, Thompson 1991). It has not only displaced both parental species but has also invaded lower parts of the tidal zone where neither parent is able to grow. Another example of adaptive evolution following hybridization is the case of Europe's weed beets (see chaps. 1 and 6). A literature review (Ellstrand and Schierenbeck 2000) found twenty-eight well-documented examples in which interspecific hybridization preceded the evolution of new lineages that became either weeds in managed ecosystems or invasives in unmanaged ecosystems. Examples involving domesticated species will be reviewed in chapter 9.

Consequences of Gene Flow versus Mutation

Mutation is defined as spontaneous genetic change. Mutation occurs either at the level of the gene, involving the change from one allele into another, or at the cytogenetic level, involving changes in chromosome structure and number. Even though mutation is the source of all genetic variation, mutation rates in plants are typically on the order of 10^{-4} to 10^{-6} per locus per generation (Grant 1975). Thus, every generation, mutation typically exerts a little "pressure" on the population that, all other things being equal, eventually results in evolutionary change (Wright 1969).

Gene flow and mutation are similar processes in that they are able to introduce genetic novelty into a population (Wright 1969). When the mutation rate (u, the fraction of new mutant alleles per locus per generation) and the gene-flow rate (m, the fraction of the population replaced per generation by immigrants) are the same order of magnitude, they are of comparable evolutionary significance.

Typical gene-flow rates that have been measured among natural plant populations are so high that gene flow is much more likely than mutation to introduce new variation into a population (Ellstrand 1992). Likewise, the typical hybridization rates measured from crop-to-wild hybridization (table 4.1) are orders of magnitude higher than typical mutation rates. Again, we expect the evolutionary impact of gene flow in these systems to be much more important than that of mutation. Even if gene flow and mutation are working in opposition, mutation is usually so infrequent as to be overwhelmed by gene flow. As stated by the great population geneticist Sewall Wright (1969), "In most cases . . . m would be so much greater than u . . . that the latter can be ignored" in population-genetic considerations.

Role of Hybrid Fitness in the Consequences of Gene Flow

Much of this chapter has focused on theory developed to predict gene-flow dynamics of single loci. But individual genes live with other genes within living organisms. One or more detrimental alleles may reside in a hybrid that is otherwise flush with heterosis. One or more beneficial alleles may reside in a hybrid that is highly sterile. How will these interactions influence whether alleles spread from domesticated plants into wild populations?

The first key consideration is the fitness of the interpopulation hybrid. A handful of experiments have now been performed to measure the fitness of domesticated-wild hybrids relative to pure wild individuals; representative examples are listed in table 4.4. Typically these experiments involve planting hybrids side by side with pure wild individuals under field conditions. The fitness correlates that are measured may include survivorship, biomass, female function (fruit and/or seed production), male function (pollen viability or ability to sire seeds), and ability to reproduce vegetatively (such as tiller production).

Relative fitness measured in such studies (table 4.4) varies tremendously. Some of the examples show enhanced fitness for the hybrids (e.g., Klinger and Ellstrand 1994). In others, fitness of the hybrid does not appear to be

Table 4.4 Some Experimental Comparisons of the Fitness of Domesticated-Wild Hybrids Relative to the Wild Parent, under Field Conditions

Cultigen	Scientific Name	Wild Relative	Some Fitness Components Measured	Hybrid Fitness Relative to Wild Parent	Citation
Bread wheat	Triticum aestivum	Aegilops cylindrica	Seed set	Much less ($\leq$1%)	Guadagnuolo et al. 2001a
Bread wheat	Triticum aestivum	Aegilops cylindrica	Seed set	Less	Snyder et al. 2000
Radish	Raphanus sativus	R. raphanistrum	Pollen fertility	Less	Panetsos and Baker 1967
Radish	Raphanus sativus	R. raphanistrum	Pollen fertility, seed set	Less	Snow et al. 2001
Radish	Raphanus sativus	Same sp.	Seed set	Greater	Klinger and Ellstrand 1994
			Ability to sire seed	No difference	Klinger and Ellstrand 1994
Rapeseed	Brassica napus	B. campestris	Survivorship	No difference	Hauser et al. 1998b (but cf. Hauser et al. 1998a)
			Seed set	Greater	Hauser et al. 1998b (but cf. Hauser et al. 1998a)
Rapeseed	Brassica napus	Hirschfeldia incana	Seed set	Much less ($\leq$1%)	Lefol et al. 1996b
Rapeseed	Brassica napus	Raphanus raphanistrum	Seed set	Much less ($\leq$1%)	Darmency et al. 1998
Rice	Oryza sativa	Same sp.	Tiller number, height	Greater	Langevin et al. 1990
Sorghum	Sorghum bicolor	S. halepense	Biomass, tiller number, seed set, pollen viability	No difference	Arriola and Ellstrand 1997
Squash	Cucurbia pepo	C. texana	Seed set	Less	Spencer and Snow 2001
			Survivorship	No difference	Spencer and Snow 2001
Sunflower	Helianthus annuus	Same sp.	Seed set	Varies	Snow et al. 1998
Sunflower	Helianthus annuus	Same sp.	Seed consumed by insects	Less	Cummings et al. 1999

significantly different from that of the wild population (e.g., Arriola and Ellstrand 1997). There are cases in which fitness is moderately depressed (e.g., Snow et al. 2001, Spencer and Snow 2001) or so depressed that plants are approaching full sterility (e.g., Lefol et al. 1996b). Indeed, in a few cases relative hybrid fitness varies with environmental conditions, sometimes enhanced and sometimes depressed compared with pure wild genotypes (Snow et al. 1998). Such variation in hybrid fitness, both within and among studies, is typical of that observed for hybrids among natural species (Arnold and Hodges 1995, Arnold 1997).

Absolute hybrid sterility is an absolute barrier to gene flow. It doesn't matter whether a hybrid holds one or one hundred alleles that could prove advantageous to the wild population. With absolute hybrid sterility, beneficial and neutral alleles will not enter a population. However, I did not find any case of hybrids between domesticated and wild plants that were absolutely seed- and pollen-sterile under field conditions.

Studies of the relative fitness of advanced-generation crop-wild hybrids are very rare. Interestingly, those on hybrids between oilseed rape *Brassica napus* and wild *B. campestris* show that fitness may change with continued introgression. The F_2s show a fitness decrease relative to the F_1 (Hauser et al. 1998a), while backcrosses of the F_1 to the wild parent show a fitness increase (Mikkelsen et al. 1996). Additionally, the fitnesses of hybrids and hybrid derivatives might be expected to vary over time and space (e.g., Snow et al. 1998, Spencer and Snow 2000).

With few exceptions, the results of the experiments summarized in table 4.4 suggest that if hybridization occurs between domesticated plants and their wild relatives, the fitness of the hybrids and hybrid derivatives may effectively alter gene-flow rates but is unlikely to stop gene flow altogether. If hybrid fitness is reduced, but not zero, gene flow is effectively reduced, but not eliminated, for neutral and beneficial alleles (Barton and Bengtsson 1986, Gavrilets and Cruzan 1998). If the fitness of the average hybrid is lower than that of the average genetically pure wild plant, the effective immigration rate of neutral and beneficial alleles would be reduced accordingly. If recombination in advanced-generation hybrids decouples these neutral or beneficial alleles from low-fitness phenotypes, they will spread through the population according to the expectations of single-gene theory described above. Conversely, alleles that occur in a heterotic hybrid have a gene-flow rate amplified proportional to the relative fitness of the hybrid. If the average hybrid fitness is higher than that for nonhybrids, the effective immigration rate of deleterious and neutral alleles would be increased proportionally. Interestingly, in the extreme case a hybrid with enhanced fitness bearing detrimental alleles expressed in later generations can result in local extinction (see Muir and Howard 1999).

Whether or not extensive introgression will occur over the generations following a hybridization event will depend on a great variety of genetic and environmental factors. While the average hybrid fitness may provide information about initial opportunities for introgression, keep in mind that hybrids often show a range of fitness. Whether introgression proceeds or not may be more a function of whether some fraction of hybrids have high male or female fitness, regardless of the average fitness of the group. As concluded by Arnold et al. (1999), "Extremely low fertility or viability of early-generation hybrids (e.g., F_1, F_2, B_1) does not necessarily prevent extensive gene flow and the establishment of new evolutionary lineages."

Linkage between genes that are experiencing different kinds of selection can also play a role in introgression. For example, theory has shown that neutral alleles tightly linked to highly advantageous alleles can be broadcast very quickly through one or more populations in a process known as a "selective sweep" (Maynard Smith and Haigh 1974, Majewski and Cohan 1999). Likewise, an allele may have complex effects, benefiting one or more life-history stages but posing costs at others (Muir and Howard 2001). The specifics of these cases will determine the eventual fates of the alleles and the populations involved.

For readers seeking more information, the theoretical and empirical work on hybrid zones reviewed by Barton and Hewitt (1985) and Arnold (1997) provides a detailed examination of complex gene-flow/selection interactions.

Part II / Caught in the Act

CHAPTER 5

Evidence for Recognizing Natural Hybrids

If it looks like a duck, walks like a duck, and quacks like a duck, then it just may be a duck. —WALTER REUTHER, LABOR LEADER

To be uncertain is uncomfortable, but to be certain is ridiculous. —CHINESE PROVERB

The first part of this book makes it clear that it is not rare for plants to hybridize, and when they do, that hybridization can have important evolutionary consequences. The second part examines the evidence for natural mating between domesticated plants and their wild relatives. But before proceeding to the case studies for the world's most important crops, it is worthwhile to review the various techniques used to amass that evidence. Each technique has its strengths and weaknesses. But as the number of tools used increases, so increases the level of confidence for assigning hybrid ancestry (Gottlieb 1972). For some crops the evidence for hybridization, and sometimes introgression as well, is quite strong; for others it is largely circumstantial, identifying situations requiring further research.

This chapter addresses the question, How do plant scientists determine whether natural hybridization has occurred? It lays out the tools most frequently used to determine whether spontaneous hybridization has occurred. Below I evaluate the strengths and limitations of the methods commonly used to judge the likelihood of natural hybridization, but my list is not exhaustive. Readers seeking a more thorough review are directed to Stace (1975b), particularly the chapter "Recognition of Hybrids," for a critical evaluation of methods other than molecular markers, and to Rieseberg and Brunsfeld (1992) for a review of the use of molecular markers for identifying hybrids and hybrid derivatives.

Intermediacy

When populations of two related species are found growing near one another, the discovery of a plant that shares some of the characteristics of both species often leads to the tentative conclusion that it is a hybrid. These intermediate characters are often morphological but could also be phenological, physiological, developmental, or the like. Intermediacy occurs in two ways. On one hand, characters that are roughly halfway between one parent and the other may suggest hybrid origins—for example, pink flowers on the putative hybrid of red-flowered and white-flowered parents. The oracle oak described in chapter 3 is a good example. On the other hand, a combination of characters of the potential parents may also suggest hybridity—say, a white-flowered, large-leafed presumed hybrid of a white-flowered, small-leafed parent and a red-flowered, large-leafed parent. Most often hybrids have some characters that are halfway between the two parents, some that are identical to those of one parent, and some that are identical to those of the other. Studies of artificially created plant hybrids typically show that about 45 percent of a hybrid's characters fall between the parents, another 45 percent are the same as in one parent or the other, and the remaining 10 percent are actually more extreme than in either parent (Rieseberg and Ellstrand 1993, Rieseberg and Carney 1998).

The case of the "Piãta" variety of coffee offers a good example (Medina Filho et al. 1995). This cultivar is a clone of a spontaneous seedling, discovered in a plantation of domesticated arabica coffee *(Coffea arabica)* growing near some wild trees. Piãta has a mixture of characters of the domesticated species and the wild species *C. dewevrei.* Its leaves are bronze-colored and have a pointed, narrowed tip, similar to *C. arabica,* but are intermediate in size between *C. arabica*'s small leaves and the large green leaves of *C. dewevrei,* which have a rounded tip. Piãta's fruits are more like those of *C. dewevrei,* larger and darker red than those of *C. arabica.* As we will see below, evidence based on other methods has supported the hypothesis that Piãta is a hybrid of *C. arabica* and *C. dewevrei* (Medina Filho et al. 1995).

When other lines of evidence are used to test the hypothesis of hybrid ancestry of an intermediate plant, hybrid ancestry is usually confirmed. Occasionally such plants are found *not* to be hybrids (several examples are listed in Rieseberg and Wendel 1993). Morphological and certain other types of intermediacy can be due to factors other than hybridity. In some cases the environment may play a role. Imagine that species A typically grows in a relatively benign habitat, and species B grows in one that is more stressful. If species A grows in that stressful environment, it may be stunted or suffer other problems that make it appear to take on the characters of species B. For example, where *Juniperus virginiana* reaches the westernmost limits of its range in Texas, the soils are poorer and the weather is drier than in its typical niche in the southeastern United States. In that location it overlaps with and superficially appears to converge with *J. asheii*, which is at the easternmost limits of its range. Those trees were presumed to be hybrids or hybrid derivatives (Hall 1952). However, analysis of species-specific secondary chemicals from the putative hybrids reveals that the plants in question are pure *J. virginiana*, and closer scrutiny of their morphology confirmed that they are not hybrids (Flake et al. 1969, Adams and Turner 1970). Thus, phenotypic plasticity alone may explain why some individuals of one species grow to resemble another.

Also, in certain environments convergent evolution may result in one species resembling another. In fact, some crop mimics are excellent examples of such convergence (reviewed by Barrett 1983). Farmers who remove weeds by hand are generally able to identify them because they look different from the crop. Weeds resembling the crop may be overlooked. Thus, selection can be quite strong for weeds to evolve to mimic the crop in which they are growing. Barnyard grass *(Echinochloa crus-galli)* is a classic example (Barrett 1983). One variety of this species, *E. crus-galli* var. *oryzicola,* is one of the worst weeds of rice *(Oryza sativa)* (Holm et al. 1977). Barnyard grass is not closely related to rice; the two cannot hybridize. Nonetheless, generations of hand-weeding have led to the evolution of the mimic *E. crus-galli* var. *oryzicola,* which is superficially much more similar to rice than to its progenitor, the nonmimic *E. crus-galli* var. *crus-galli.* In fact, they are so similar that the barnyard grass variety *oryzicola* is very difficult to distinguish from rice before they flower. It is certainly possible that apparent hybrids between a crop and a wild relative could instead be crop mimics.

It is also possible that crop mimics could be true hybrid derivatives that evolve under the joint effects of gene flow and selection. Traits received from recurrent backcrossing to the crop parent might make a hybrid weed appear so similar to the crop as to frustrate control efforts. To overcome the problem of distinguishing between wild and cultivated seedlings of rice, Indian plant breeders developed purple-leafed rice cultivars that could be easily distinguished from the green-leafed wild forms. Shortly after their introduction, natural hybridization between crop and weed, followed by selection by hand-weeding favoring purple-leafed morphs, resulted in the establishment of purple-leafed weeds (Dave 1943, Oka and Chang 1959, Harlan et al. 1973, Parker and Dean 1976). (Another example of the evolution of crop mimicry via hybridization is Europe's weed beet; see chaps. 1 and 6.) Thus, the evolution of mimetic weeds can occur with or without a history of hybridization.

Therefore, a plant that is morphologically intermediate between a domesticated plant and a wild relative may or may not be a hybrid or hybrid derivative, and the presence of both possible parents adds additional support for hybridity (Stace 1975b). But without further supporting evidence, the alternatives of phenotypic plasticity or nonhybrid crop mimicry are also possible (National Academy of Sciences 1989). Conversely, a lack of morphological intermediacy is not proof of a lack of hybrid ancestry. Many generations of backcrossing to one species may result in the loss of most, but not necessarily all, of the alleles from the other species that contributed to the lineage. Indeed, crop breeders have successfully employed repeated backcrossing to move desirable alleles from a wild species into a number of crops (Fehr 1987). Clearly, it is worthwhile to use more criteria than intermediacy alone to determine whether or not a plant is a natural hybrid.

Sterility

Plant hybrids, including hybrids between crops and their wild relatives, often show some decrease in seed or pollen fertility relative to their parents (see the examples in table 4.4 and Stace 1975b). Full or partial sterility is often used as supporting evidence that a plant is a hybrid (Stace 1975b). The fertility of pollen may be tested directly in terms of its ability to sire seeds or indirectly with pollen grain counts, pollen viability stains, or inspection of pollen grains to see if they are aborted or misshapen. Seed fertility may be assessed directly by counting and germinating the seed set by a plant or indirectly with seed viability stains. Of course, variation in pollen and seed fertility may occur in natural populations as well. Therefore, it is important for studies using fertility as a test of hybridity to measure the fertility of the presumed parents as a control.

Not all hybrids have decreased fertility. It is not rare for plant hybrids to have the same fitness as the parents or even higher fitness (Arnold 1997, Arnold and Hodges 1995). The same is true for hybrids between crops and their wild relatives (see the examples in table 4.4). Therefore, the lack of full or partial sterility is not proof of a lack of hybrid ancestry. Sterility can contribute to the assignment of hybridity, but additional criteria are helpful in ascertaining higher levels of confidence in that assignment.

Evidence with a Known Genetic Basis

Hybridity is often assigned to plants observed to have some intermediacy and perhaps some sterility that are growing in the vicinity of both parents. And indeed, when such cases are further investigated the additional evidence usually supports the hypothesis of a hybrid origin for these plants. Nonetheless, without some genetic evidence of hybridity, the alternatives loom large.

Characters with a known genetic basis argue a much stronger case. If suspected hybrids or hybrid derivatives have a combination of genetic characteristics that are specific to each parental type, then hybridity is supported. However, such data are still not necessarily unambiguous confirmation of hybridity.

Imagine the following scenario. An ancient species is polymorphic at one of its loci for allele A and allele B; the allele frequencies are 50 percent each. That same species gives rise to two species; let's call them species 1 and species 2. Further assume that species 1 is fixed for allele A, and that species 2 has allele B at a frequency of 97 percent and its lineage has retained allele A as well, at a frequency of 3 percent. Individuals that retain a primitive character-state represent a situation known as "lineage sorting" or "symplesiomorphy" (discussed in Rieseberg 1997; see fig. 5.1). A scientist who encounters individuals of species 2 that carry allele A ("symplesiomorphs") might presume that those individuals are derivatives of hybrids between species 1 and species 2, when in reality they simply bear a rare allele retained during the evolution of the species. As a result, it is much easier to reject the hypothesis of hybrid origin than to confirm it in descriptive studies with one genetically based marker. The use

A.

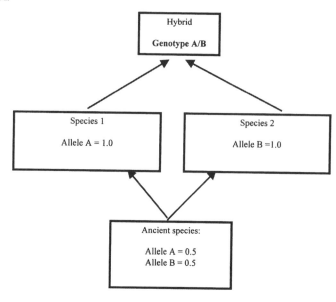

B.

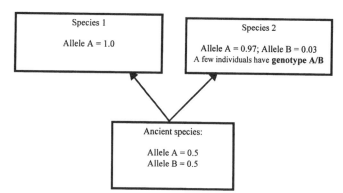

Fig. 5.1 "Lineage sorting" or "symplesiomorphy" can lead to misassignment of hybridity. Species 1 and species 2 are both descended from a common, polymorphic ancestor. In situation A, species 1 is fixed for one allele; species 2 is fixed for the alternate allele. Their hybrids have the heterozygous genotype A/B. In situation B, species 1 is fixed for one allele; the alternate allele is very common in species 2, but species 2 retains the other allele in a low frequency. Individuals of species 2 that are heterozygous for that rare allele might be inappropriately deemed to be hybrids of species 1 and species 2 or to have an ancestry that involves hybridization between species 1 and species 2.

of more than one locus or linked markers greatly increases the probability of distinguishing symplesiomorphs from hybrid derivatives (see Linder et al. 1998 for an exemplary study). Likewise, experimental approaches may circumvent this problem.

Experimental Approaches

Artificial Synthesis of Hybrids

As noted above, the typical plant hybrid has some characters that are specific to one parent, other characters that are specific to the other parent, and a small number that are more extreme than either parent. In most cases it would be impossible, a priori, to guess how morphological characters would be expressed in a hybrid. Therefore, if the cross can be made by hand, the product can be compared with the plant in the field.

In the case of the cultivated oat, *Avena sativa,* and the wild oat, *A. fatua,* crosses have been made with both species as seed parents. The first-generation hybrids tend to resemble the cultivated parent in most characters (Barr and Tasker 1992, Stace 1975a). Segregation of traits in later generations creates plants that closely resemble those identified as spontaneous hybrids in the field, suggesting that the F_1 plants are often overlooked.

While artificial synthesis may be a good method for identifying first-generation, and sometimes advanced-generation, hybrids, it is not necessarily an easy one to accomplish. Barriers to hybridization may be challenging. A classic example is *Brassica napus,* an important vegetable and oil crop. Plenty of evidence is available demonstrating that this species is an allotetraploid derivative of a hybrid of *B. oleracea* and *B. rapa,* but "the parental species are extremely difficult to cross" (McNaughton 1995a). Nonetheless, "the hybrid has been produced experimentally. It resembles a weak *B. napus,* but is completely sterile. . . . By chromosome doubling, fertile *B. napus* can be synthesized" (Harberd 1975).

It is possible, but extremely unlikely, that symplesiomorphs, nonhybrid crop mimics, or plastic individuals might have precisely the same combination of characters as a true hybrid. Thus, this method is a good one for testing hybridity. It is surprising that it is used so rarely. One explanation may be the difficulty of creating hybrids. Another may be the fact that plant hybrids are often identified solely by their reproductive characters, which might take years to observe in a perennial species with delayed maturity.

F_2 Segregation

If plants are true F_1 hybrids, then Mendel's laws dictate that both parental types might be recovered in the F_2 generation. Progeny segregation studies are sometimes used to support the hypothesis of hybridity (Stace 1975b). Indeed, incorrectly assigned "hybrids" resulting from symplesiomorphs, phenotypic plasticity, or nonhybrid crop mimicry should be recognized through this method because they would all breed true for their characters without segregation.

This method has sometimes been used successfully to confirm the origins of hybrids between crops and their wild relatives. The spontaneous hybrid Piãta coffee cultivar, discussed above, is an example. Its seeds segregate for the endosperm colors that characterize its two putative parents, *Coffea arabica* and *C. dewevrei* (Medina Filho et al. 1995).

F_2 segregation can be a powerful method for identifying first-generation, and sometimes advanced-generation, hybrids, but it is not necessarily easily accomplished. As pointed out above, some hybrids have high levels of sterility; thus, F_2s may be difficult to obtain. Long-lived plants, such as trees, may take years to become sexually mature.

Even if enough progeny are produced, "the chances of recovering exact parental types are very low; . . . one should merely expect to obtain near-parental types" (Stace 1975b) because recombination will decouple linkage relationships. That is, unlinked genes will assort randomly from one another in second-generation hybrids so that the traits from one of the parents will not be inherited as a block but will tend to assort themselves randomly among the progeny. Another problem is that Mendelian ratios are often distorted in interspecific hybrids (Zamir and Tadmor 1986) such that alleles from one or the other parental species may be overrepresented in the F_2s.

Experiments Measuring Hybridization Rates in the Field

Experiments are sometimes performed to measure whether spontaneous hybridization can occur under field conditions. Typically these experiments involve a stand of crop plants intermixed with or surrounded by synthetic populations of a wild relative. The plants are chosen so that the crop bears an allele that is absent in the wild plants. Progeny testing of seed harvested from the wild plants or seedlings grown from that seed reveals whether the crop was their male parent. Some representative examples are summarized in table 4.1.

One study of this type was conducted by Klinger et al. (1991) to determine whether wild radish and cultivated radish (both *Raphanus sativus*) naturally hybridize under field conditions in California. The authors grew the cultigen in the same fashion as might a farmer who was growing the crop for seed production (radish grown as a vegetable is typically harvested before it flowers). They surrounded the crop with small synthetic populations of wild radishes, just as such populations might occur naturally. After both the cultivated type and the wild plants had flowered, they harvested mature seed from the wild plants. The cultivar was homozygous for one isozyme allele; the wild experimental plants were fixed for other alleles at that locus. Thus, when progeny from the wild plants were screened, progeny bearing the crop allele would have to have been sired by the cultivar. Progeny testing revealed that the crop readily hybridized with wild radish, at extremely high rates at the distance of one meter and lower rates at greater distances, but hybridization was still detected at the most distant stations, one thousand meters away (see the summary in table 4.1).

Highly controlled experiments of this kind are useful for demonstrating whether a crop will spontaneously mate with a wild relative in the field. Problems associated with plasticity, mimicry, and symplesiomorphy are avoided. Nonetheless, experiments, by their very nature, are somewhat artificial. They demonstrate whether or not natural hybridization can occur in a given location under the experimental conditions, but they do not unambiguously confirm that suspected hybrids in the field are indeed hybrids.

Descriptive Studies of Genetically Based Markers

Next to cases of intermediacy, descriptive studies employing genetically based markers make up the bulk of research on natural hybrids between domesticated plants and wild relatives. Few morphological markers are reliable enough or genetically defined enough to serve this purpose (but see fig. 5.2). Therefore, laboratory-based analyses are often necessary. Plants with presumed hybrid ancestry are collected and brought back to the lab where they are subjected to analysis for characters with a known genetic basis. These characters are then compared with a sample of control plants from the two putative parents that are presumably genetically "pure," that is, those without a

Fig. 5.2 Infructescences and fruits of teosinte *(Zea mays* ssp. *mexicana),* maize *(Z. mays* ssp. *mays),* and their F$_1$ hybrid. An example of genetically based differences (Doebley 1996) as expressed in parents and hybrids. *Left to right:* Teosinte fruits, teosinte infructescence, hybrid infructescence, maize infructescence. Photo by Janet Clegg. Used with permission.

history of hybridization. If presumed hybrids have an allele diagnostic of each parent, then hybridity is supported. As noted above, such evidence works well for refuting plasticity or nonhybrid convergence, but not necessarily symplesiomorphy.

Chromosomes (Cytogenetic Markers)

Closely related species often differ in chromosome number and/or chromosome structure. Such cytogenetic differences are the case for many domesticated plants and their closest relatives. In particular, certain crops appear to be polyploid derivatives of diploid wild species with twice the number of chromosomes. Crops with polyploid chromosome numbers relative to their closest relatives include durum wheat, bread wheat, strawberry, peanut, sweet potato, banana, white potato, taro, arabica coffee, and tobacco (Hancock 1992, Simpson and Ogorzaly 2001).

If a plant has a combination of chromosomal characters specific to the parental taxa, that is usually strong support for hybridity. The spontaneous hybrid coffee cultivar Piãta, discussed above, is an example. Its chromosome number is forty-four. Cytogenetic inspection of those chromosomes suggests that half of them resemble those of *Coffea arabica* and the other half resemble those of *C. dewevrei* (Medina Filho et al. 1995). The interpretation of those data is that Piãta was probably a natural hybrid from the fusion of a normal gamete (n = 22) of *C. arabica* (2n = 44) and an unreduced gamete (2n = 22) of *C. dewevrei*.

Cytogenetic data work well to refute cases of plasticity or nonhybrid mimicry, but they have their limitations. A cytogenetic approach may be unavailable if there are no discernable differences in chromosome structure or number between the parental taxa. If there is natural chromosomal variation within one or both parental species in question, the possibility of symplesiomorphy may be hard to refute. Also, generations of backcrossing and introgression may have selected individuals with chromosomal number and structure much like one parent, while still holding some alleles from the other parent such that a lack of cytogenetic evidence may not indicate a lack of hybrid ancestry.

Isozymes (Allozymes)

Isozymes represent the most common class of genetic markers for descriptive studies of putative domesticate-wild hybridity as well as for experimental studies to measure natural domestic-to-wild hybridization rates. Isozymes are the enzyme products of genes that are identified by first separating them in a gel using an electric field (electrophoresis) and then staining for the activity of specific enzymes (Weeden and Wendel 1989). The results are stained bands of different mobilities. This technique is now over four decades old, and the genetic basis of the banding patterns of dozens of plant enzyme loci is now well understood (Wendel and Weeden 1989). Generally, allelic products are expressed co-dominantly (i.e., simultaneously) so that homozygotes produce one band and heterozygotes produce multiple bands. (Although the term "isozymes" first described differentially migrating enzymes, and the term "allozymes" originally described isozymes whose genetic basis was known, the two are now used interchangeably, along with the term "isoenzymes.")

For an example of isozymes in action, let's return to Piãta coffee one last time. Despite all of the evidence supporting the origin of this cultivar as a hybrid of *Coffea arabica* and *C. dewevrei*, additional support was sought from isozyme analysis (Medina Filho et al. 1995). The alleles of the genes coding for the enzymes phosphoglucoisomerase, phosphoglucomutase, and alcohol de-hydrogenase were found to be specific to each of the presumed parental species. Those alleles were found segregating in the seeds of Piãta, supporting its interspecific origin. With all of this evidence, the hybrid origin of Piãta seems very likely indeed.

Isozymes have many strengths relative to morphological markers. Their genetic basis is well understood. It is not unusual for ten or more loci to be surveyed simultaneously from a sample. Because of co-dominance, the precise allelic constitution of an individual can be scored for each locus. The presence of loci with alleles diagnostic for two different species can easily refute convergent evolution without hybridization or plasticity. As the number of loci with diagnostic alleles increases, the alternative of symplesiomorphy decreases. But certain DNA-based markers offer a greater number of loci to sample and a greater opportunity to refute symplesiomorphy.

Nonetheless, isozymes have certain strengths compared with DNA-based markers as well. Banding patterns for a given enzyme are generally easily interpretable for all seed plants. If an isozyme locus is assigned to one member of a genus, the homologous locus usually can be identified in another member. Isozyme mutation rates are high enough that it is not unusual to detect three or more alleles at polymorphic loci, but low enough that populations rarely evolve convergent alleles, if ever. Null (recessive) alleles are very rare; co-dominance is standard. In 1992 Rieseberg and Brunsfeld stated that "although isozymes have been used effectively in a number of studies of introgression in plants . . . , it seems likely that they will be supplemented and possibly replaced by [DNA-based] nuclear RFLP [restriction length fragment polymorphism] markers in future studies." Now, a decade later, isozymes are still the predominant markers used for descriptive and experimental studies of plant hybridization and introgression. No single DNA-based technique has yet become more popular than isozymes as a tool for identifying natural hybridization.

Generally, if the taxa thought to be hybridizing are genetically well differentiated, an initial survey with isozymes is the best first step for testing the hypothesis of interspecific hybridization and for characterizing whether any introgression has occurred. DNA-based markers are necessary for thorough exploration of the effects of hybridization at the genomic level. A number of reviews compare the strengths and weaknesses of isozymes with DNA markers in detail (e.g., Ouborg et al. 1999, Parker et al. 1998, Rieseberg and Ellstrand 1993).

DNA-Based Markers

A heterogeneous set of techniques are available to detect DNA variation directly. In terms of identifying hybrids or plants with hybrid ancestry, each of these techniques has its own strengths and weaknesses. For a detailed comparison, see one of the many reviews (e.g., Avise 1994, Ouborg et al. 1999, Parker et al. 1998, Rieseberg and Brunsfeld 1992, Rieseberg and Ellstrand 1993, Whitkus et al. 1994, Wolfe and Liston 1998). The following mini-review is abstracted from those reports. Readers seeking more thorough information on these and other methods are encouraged to consult the above-mentioned reviews or the website "PCR Techniques" by Seranne Howis (http://rucus.ru.ac. za/~wolfman/Essays/PCR1.html), which explains and compares the extant techniques.

As a group, DNA-based markers offer the opportunity to inspect more loci and more alleles than isozymes. They are used to scrutinize both nuclear genomes as well as organellar (chloroplast and mitochondrial) genomes. Nuclear genes are inherited in the standard, biparental Mendelian fashion. DNA in the chloroplast and mitochondrial organelles is a circular molecule of several genes inherited directly (clonally) from parent to progeny. Organellar genomes are often transmitted uniparentally (but this varies to some extent among plant families) and without recombination from maternal parent to progeny. Chloroplast DNA is highly conserved with low polymorphism within subspecies or species, but differences may occur between species. Both chloroplast and mitochondrial DNA are typically analyzed by isolation and subsequent RFLP analysis (see below) in the same way that RFLP analysis is applied to nuclear genes. Because they contain multiple, nonrecombinant genes, cytoplasmic genome analysis greatly increases the potential for the simultaneous scoring of multiple markers in a real hybrid or hybrid derivative. Such data make the alternatives of symplesiomorphy or nonhybrid convergence extremely unlikely. Introgression of organellar genomes ("cytoplasmic introgression") and its consequences in plants have been the subject of an extensive review (Rieseberg and Soltis 1991).

Of the many classes of DNA-based markers, three have made substantial contributions to the recognition of plant hybrids. They are restriction length fragment polymorphisms (RFLPs), random amplified polymorphic DNA (RAPD, and a very similar technique, inter-simple sequence repeats, ISSRs), and amplified fragment length polymorphisms (AFLPs). I briefly review each of these below.

The first DNA-based markers to be developed were RFLPs. RFLPs are detected by cleaving ("digesting") genomic DNA with restriction enzymes, separating the products on a gel, and probing for a specific DNA fragment. Probes used in RFLP analysis can be generated from cloned gene encoding sequences, or from nuclear genomic, chloroplast DNA or mitochondrial DNA fragments, or from DNA segments amplified using polymerase chain reaction (PCR, explained below). Thus, depending on the probe used, RFLPs can be used to analyze mitochondrial DNA variation, chloroplast DNA variation, or nuclear variation, ranging from the multigene family of ribosomal DNA to a single locus. RFLP variation may be the result of length mutation and/or point mutation at a restriction enzyme cleavage site at a given chromosomal location. The typical RFLP locus is co-dominant and has more alleles than an isozyme locus.

RAPD markers are based on the PCR amplification of random locations in a plant genome. PCR is a method that repeatedly doubles ("amplifies") a tiny amount of a DNA sequence until the concentration of that particular sequence is so high that it can be resolved with a DNA-specific stain. Typically, a short stretch of nucleotides are used as a primer to both find and attach to ("anneal") a specific DNA sequence and serves as a starting point for amplification. For RAPDs a single oligonucleotide is used to prime the amplification of genomic DNA. Because these primers are about ten nucleotides long, they have the possibility of annealing at a number of locations in the genome. For amplification products to occur, the binding must be to sequences that are inverted relative to each other and about 150–4,000 base pairs apart. The amplified products are separated on a gel, and the DNA is stained directly without a probe. Many loci can be screened by this method. The amplified RAPD loci are scored as the presence or absence of a dominant allele. Heterozygotes usually cannot be distinguished from individuals that are homozygous for the dominant allele. RAPDs whose genetic basis is demonstrated through Mendelian analysis offer the opportunity to screen up to hundreds of loci (for an example of RAPDs used for identifying plant hybrids, see Kuehn et al. 1999). The ability to produce interpretable and reproducible

RAPDs has improved considerably since the technique was initially developed, but RFLPs and AFLPs are still considered more reliable (Parker et al. 1998). ISSRs are markers derived from techniques nearly identical to RAPD techniques except that ISSR primer sequences are designed from microsatellite regions that are long stretches of highly repetitive, noncoding DNA. See Wolfe et al. (1998) for an example in which ISSRs were used to identify hybrid ancestry as well as a review of their potential for studies of natural populations.

AFLPs combine the characteristics of RFLPs and RAPDs. They are molecular markers generated by a combination of restriction digestion and PCR amplification. After restriction digestion, "adaptor" sequences are attached to the ends of the restriction fragments. The product is thousands of adapted fragments. Primers composed of the adaptor sequence plus one to three extra nucleotides are used to selectively amplify a subset of those fragments with PCR. The amplified fragments are separated on a gel, and the DNA is scored directly. The technique yields dominant alleles for a high number of loci. Interested readers seeking more detailed information should consult a recent review on AFLPs (Mueller and Wolfenbarger 1999).

Conclusions

The two quotations that appear at the beginning of this chapter characterize the tension in assigning a hybrid origin to an individual. For hybrid identification, like most science, refuting a hypothesis is much easier than proving a hypothesis—if indeed unambiguous proof is ever possible. As Gottlieb (1972) pointed out, the best one can hope for is the increased confidence of assigning hybridity as data accumulate.

In the study of hybridization between domesticated plants and their wild relatives, a vast majority of analyses of intermediate plants growing in the presence of their putative parents have resulted in the support of the hypothesis of hybridity. Indeed, among the hundreds of articles I have inspected on natural domesticate-wild mating, I have found only one study refuting the hypothesis of possible hybridization (Lowe et al. 2000). Therefore, for the purposes of this book I proceed with the practical view of Walter Reuther, but keeping in mind the Chinese proverb. I operate under the assumptions that (1) intermediates between cultigens and wild plants are likely to be of hybrid origin and (2) if any further genetically based evidence is offered, that "confirms" hybridity—while recognizing that to be certain of anything is ridiculous.

Now that we have gained an understanding of the tools genetic detectives can use to identify parents and ancestors, we are ready to return in the next chapter to "The Case of the Bolting Beets" and see how those methods were used to solve that mystery.

CHAPTER 6

The Case of the Bolting Beets
Part II. Detectives Arrive at the Scene

There are no illegitimate children, only illegitimate parents.
—EDNA GLADNEY, PHILANTHROPIST

Weed beets had appeared mysteriously in the sugar beet fields of Europe in the latter third of the last century. At first the reports were scattered, but eventually it became clear that sugar beet growers had a new weed on their hands, one that mimicked the crop closely until the plants flowered. If weed beets weren't promptly removed or killed, the seed they set flooded the seed bank with even greater numbers of weeds. It was time to figure out how Europe's weed beets had evolved.

In the early 1990s the detectives got to work identifying the parentage of Europe's weed beets. Various research groups genetically analyzed the bolting weed beets and compared them with nonbolting cultivars as well as with the wild, but nonweedy, sea beet. Identifying the source of weed beets would provide crucial information for the prevention and management of the scourge of Europe's sugar industry. Were the new weeds rogue cultivars, wild sea beets invading a new habitat, or a cultivar-wild hybrid?

The first question asked about these children of uncertain parentage was, How did they obtain a gene for bolting? Sugar beets don't have the allele. It is always possible that a mutation could give rise to a new bolting allele. But the allele for bolting (*B* allele) is generally dominant (Boudry et al 1994b). Thus, if such a mutation occurred in lines grown for seed production or in a breeding program, the allele would be eliminated under a watchful breeder's eye. It would be impossible for so many mutants to escape, unless perhaps some environmental trigger for expression was present in northern Europe's sugar beet production fields that was absent in southern Europe's seed production areas.

What's the distribution of the bolting allele in wild populations? The *B* allele is common in wild sea beets of the Mediterranean but totally absent in those populations along the English Channel and North Sea (Van Dijk and Boudry 1992). This geographic pattern is an important clue because wild sea beet populations growing closest to where weed beets first appeared, that is, sugar beet production regions, do *not* have the bolting allele. In contrast, those sea beets growing closest to the sugar beet seed multiplication areas not only have the allele but have it in abundance. This pattern alone suggested that the weed beets were more likely to be hybrids created by illicit sex in the south of France than to be wild sea beets from the north coasts that had somehow managed to invade sugar beet plantations. However, this argument was based only on one line of evidence from a single gene. It was still possible that an extremely rare mutant sea beet from the North Sea coast had found its way into the sugar beet fields. The hunt for supporting evidence intensified.

Not surprisingly, the first corroborating data came from the country hardest hit by the weeds: France, the world's foremost producer of sugar beets (FAOSTAT Statistics Database, http://apps.fao.org/). Santoni and Bervillé (1992) sampled bolters from the vicinity of sugar beet seed production fields as well as from those appearing in sugar beet production fields. They conducted RFLP analysis on the nuclear-based ribosomal DNA genes of these plants (see chap. 5 for a description of RFLP analysis and other genetic analysis techniques discussed in this chapter). They compared the genotypes they obtained from bolters with genotypes from control populations of pure sugar beet and pure wild sea beet. A certain allele, designated "V-11-2.9," was found in every cultivated sugar beet tested, and it was not present in any plants from pure sea beet populations. Almost all of the sampled bolters held at least one copy of the V-11-2.9 allele, indicating that the vast majority had sugar beet as a parent or ancestor. This pattern was true for twelve of the thirteen bolters sampled from the vicinity of the seed production field as well as thirty-five of thirty-six bolters sampled from five different sugar beet production areas in Europe.

All eight bolters sampled from a sugar beet production field in eastern France were heterozygous for an allele from wild sea beet not found in any of the sugar beets sampled (designated "V-10.4-2.3"). These data support the contention that the bolters were hybrids. Likewise, most of the bolters sampled from southwestern France appeared to be hybrids as well.

The evidence for hybrid ancestry was not as clear in the other populations. Only one-quarter of the bolters sampled from western France and Italy were heterozygous for a wild sea beet allele. In these locations it was possible that some bolters had hybrid ancestry and the others were descended from sugar beet mutants. Alternatively, all of the plants could have had hybrid ancestry, and the cultivar marker allele could have been lost in advanced-generation hybrids through segregation. In the sample from Yugoslavia, not one bolter of the eight analyzed had an allele specific to the wild plants, suggesting that all these plants were pure sugar beets or were descended from the cultivated type without a history of hybridization. However, it is still impossible to exclude the possibility of hybridization and subsequent loss of the marker through segregation in this small sample. Clearly, while hybridization seems to have played a role in the origin of bolters from the field in eastern France, the bolting phenotype and one additional genetic marker were insufficient to tell the whole story.

Pierre Boudry and his colleagues (1993, 1994a) conducted a more comprehensive study. They first assembled a greater variety of beets for genetic analysis. Representing their cultivated material were thirty-five sugar beet varieties and four varieties of table beets. Their wild collections included an array of forty-three wild sea beet populations from the coasts of western Europe (from the Dutch North Sea to the French Mediterranean), three wild beet populations from inland areas in southwestern France, and nine populations of weed beets collected directly from sugar beet fields in northeastern France. In all, about two hundred plants were sampled for seed. The authors grew plants from these seed collections in the greenhouse. To identify the maternal origin of the weed beets, they analyzed genes that beets inherit directly from their maternal parent: those of the mitochondrion and the chloroplast. Sugar beets typically have a combination of mitochondrial and chloroplast genotypes designated "CMS" to signify the mitochrondrial and chloroplast genotype inherited from their "cytoplasmic male-sterile" seed parent (see chap. 1) (Saumitou-Laprade et al. 1993). Genetic analysis via RFLPs showed that the CMS combination is absent from the wild sea beet populations (Cuguen et al.

1994, Forcioli et al. 1994). If the maternal ancestor of the weed beets was sugar beet, then we would expect to find the cultivated beet's mitochondrial and chloroplast genotypes directly passed down from maternal plant to seed. Indeed, the CMS combination predominated in about 90 percent of the bolting weed beets of northeastern France (Boudry et al. 1994a) and occurred in about 25 percent of the wild inland plants of southwestern France. Clearly, the maternal ancestor of France's weed beets is sugar beet.

What about the paternal history of the weed beets? Boudry et al. (1994a) used six polymorphic nuclear RFLP loci to genetically characterize their samples of cultivated, wild, and weedy beets, but they were unable to find alleles that were specific to the cultivars. They were, however, able to calculate a set of genetic distances (genetic distance is a quantitative estimate of genetic affinity based on allele frequencies; see Nei 1978) for their samples. They found that the genetic constitution of both the wild inland beets of southwestern France and the weed beets of northeastern France were intermediate to the cultivated beets and the wild beets of the Mediterranean. In fact, the weed beets of northeastern France were genetically closer to the cultivated varieties and the Mediterranean wild beets than they were to nearby populations of wild beets of France's north coast. The data specifically point to the cultivated beet and the wild inland sea beets of southwestern France as the most likely parents of the weed beets of northern France. The same research group (Desplanque et al. 1999) subsequently conducted a study involving a larger sample of individuals and populations; the patterns were found to be the same.

Boudry et al. (1994a) concluded, "Weed beet populations are likely to have originated from accidental pollination of cultivated beets by . . . [wild] annual beets in seed production areas. These hybrid seeds—carrying the CMS cytotype and the *B* allele—were transported and sown in northern sugar production areas. They were able to bolt, flower, and set seed after a few months in sugar beet fields due to the dominance of the annual habit (allele *B*). Short crop rotations and use of chemical herbicides instead of mechanical or manual weeding allowed the maintenance and the development of weed beets via a seed bank." The authors pointed out that "the first commercial variety carrying the CMS cytotype was released in Europe in 1969," which is concordant with the fact that weed beets had become a well-established problem on that continent by the mid-1970s. Subsequent genetic analysis (Viard et al. 2002) has confirmed that the weedy bolters that appear in rows of planted sugar beet seed are crop-wild hybrids unintentionally created by the production of sugar beet seed and that weedy bolters that appear spontaneously between the planted rows are the descendants of such hybrids.

Southern France isn't the only region of unintentional hybrid weed beet production. Another important source of European sugar beet seed is northeastern Italy's Po Valley, not far from wild sea beet populations along the Adriatic Sea. This production area provides sugar beet seed mostly for Italy and Germany, but also some for France. Mücher et al. (2000) compared the genetic characteristics of seedlings grown from weed beets from sugar beet fields of northeastern Italy and western Germany with seedlings of pure sugar beets and pure sea beets from northeastern Italy's Adriatic coast. They used a battery of genetic markers (over one hundred RAPD markers in addition to genetically based traits: bolting, hypocotyl color, growth habit, and the number of seeds per seed cluster). "Most of the bolters had a red hypocotyl and multigerm seed clusters (characteristics of sea beet), but shared the orthotropic growth habit of sugar beet." Similarly, genetic relationships sorted by a quantitative analysis based on the RAPD data placed weed beet between sea beet and cultivated beet. The authors concluded that long-distance romances between wild sea beets of the Adriatic coast and sugar beets grown for seed production in northeastern Italy apparently generate weed problems in Italy and Germany in much the same way that Mediterranean sea beets and sugar beets are generating weed beets in France.

Wild relatives other than sea beet may hybridize with cultivated beets. *Beta macrocarpa* is a native of the coastal Mediterranean region from the Near East west to the Atlantic coasts of Portugal and the Canary Islands. It is also an important weed of sugar beets and certain other vegetable crops in California's Imperial Valley. McFarlane (1975) reported plants that had the appearance of *B. macrocarpa* × sugar beet hybrids in one limited area in this region. Subsequently, Bartsch and Ellstrand (1999) used isozymes to compare the genetic makeup of *B. macrocarpa* from the Imperial Valley with that of pure populations having a history of isolation from sugar beet. That sample included plants from *B. macrocarpa*'s native range in the Old World as well as from parts of California far from beet production areas. The authors found that about 2 percent of the *B. macrocarpa* plants sampled from the Imperial Valley had alleles that were specific to sugar beet, supporting a history of hybridization between the crop and the weed. Interestingly, in contrast to McFarlane's earlier observations, these hybrid derivatives were discovered scattered in a number of locations and were visually indistinguishable from other weeds. Their isozyme profiles were those of plants with hybrid ancestry; they were not F_1s. It is also interesting that the Imperial Valley is *not* a region of sugar beet seed production. Thus, for hybridization to occur between the wild species and the cultigen, the sugar beet parents had to have been spontaneous bolters or plants that were never harvested and subsequently flowered. While hybridization between the crop and a wild relative has clearly occurred in the Imperial Valley, it is not clear whether that history of hybridization has contributed to the fact that *B. macrocarpa* has become an important agricultural weed in that region.

Indeed, hybridization between cultivated and wild beets does not always lead to increased weediness. Bartsch and colleagues (Bartsch and Schmidt 1997, Bartsch and Brand 1998, Bartsch et al. 1999) have compared the genetic constitution of wild beet populations of the Adriatic coast near the Italian beet seed production areas with pure cultivars and wild beet populations of the Adriatic coast distant from the Italian seed production region. Their dataset included both an array of genetically based morphological traits (Bartsch and Schmidt 1997) and eleven isozyme loci (Bartsch et al. 1999). In both cases the authors found cultivar-specific markers in wild populations adjacent to sugar beet seed production areas, but those markers were absent from the more distant wild populations. Those wild coastal populations that have received gene flow from the nearby cultivated beets do not appear to be weedier than those that are genetically pure.

In conclusion, it appears that the sudden appearance of Europe's weed beets can be best explained by the unexpected intrusion of wild beet genes into cultivated beet seed in both France and Italy. Does this wrap up "The Case of the Bolting Beets"? Not quite. If beet cultivar seed production continues in regions that are not isolated from wild beets, bolters will continue to contaminate much of Europe's commercial seed. If sound weed management is not practiced, weed beets will multiply through Europe's sugar beet production regions.

And there is a new concern. Sugar beets are now the object of genetic engineering. In the United States alone, two transgenic sugar beets have been deregulated and over one hundred field-test permits and notifications for genetically engineered sugar beet have been filed (Information Systems for Biotechnology database, http://www.isb.vt.edu/). Some scientists have worried that if genetically engineered sugar beets hybridize with the wild beets, the engineered genes entering the weed populations might make those weeds even more difficult to control (Boudry et al. 1994b). Indeed, concern about the possibility of engineered genes moving into weed populations was the initial motivation for much of the detective work described in this chapter (e.g., Boudry et al. 1993, 1994a, Santoni and Berville 1992). But how likely is it that engineered genes would make a difference if they appeared in weed beets? That question is the topic of the final chapter of "The Case of the Bolting Beets."

Eager to read the conclusion of "The Case of the Bolting Beets"? Complete the story in chapter 10. Otherwise, proceed to chapter 7.

Do Important Crops Mate with Wild Relatives?

Most crops have no interbreeding relatives in much of the world.

—JONATHAN GRESSEL, 1999

It is clear that spontaneous hybridization and introgression of genes from domesticated plants into wild relatives is a common characteristic of domesticated plants. —NORMAN C. ELLSTRAND ET AL., 1999

Is "The Case of the Bolting Beets" the exception or the rule for most important crops? Is mating with wild relatives a common feature of domesticated plants? How strong is the evidence? If hybridization does occur, does it appear to be common or rare? Do crop genes appear to persist in wild populations for generations after initial hybridization, or does hybrid sterility prevent subsequent introgression?

In this chapter I review what is known about natural hybridization between the world's twenty-five most important food crops and their wild relatives. These case studies provide examples of the degree and extent of spontaneous hybridization and subsequent introgression between crops and their wild relatives. This diverse sample also reveals the generality of the phenomenon.

The treatment of each crop is presented as a freestanding unit for the reader whose motivation is to learn about one or a few crops. However, such readers would be wise to read chapter 5, "Evidence for Recognizing Natural Hybrids." Also, such readers should be aware that my reviews are not exhaustive. While I have attempted a thorough review of each crop, it is beyond the scope of this book to catalogue and evaluate every study of spontaneous crop hybridization with every cross-compatible wild relative.

The twenty-five most important crops grown for human consumption (whether directly or after processing) (table 7.1) are a suitable sample for assessing whether hybridization is a general feature of all domesticated plants. To capture crops of global importance, I chose to review the most important plants in terms of area planted, instead of economic value or total yield, to avoid idiosyncratic high-value or high-yield crops that might be restricted to a very narrow geographic region. The most widely grown crops have such importance that they are among the best studied of all domesticated plants. In fact, some of them (e.g., maize) are probably among the best studied of all organisms.

I used the FAOSTAT Statistics Database (http://apps.fao.org/) to obtain the most recent (2000) estimates of area harvested for each crop. Because some crops comprise more than one species, I reviewed as many species as necessary to account for a substantial majority of the area harvested. Collectively, these crops were grown on almost ten million square kilometers in 2000, somewhat more than the entire landmass of the United States.

As a group, the case studies represent a relatively heterogeneous group of thirty-one tropical, subtropical, and temperate species. Even though cereals make up a substantial minority (12) of the species reviewed, the collective list represents twenty-six genera from twelve plant families, including annuals and perennials, herbaceous plants and woody ones, crops with a great variety of uses (e.g., maize and coconut) and those with but a single product (coffee). Even though I restricted my focus to species used for human consumption, the same species often have other uses as well, such as fiber or animal food. The species vary dramatically in their reproductive biology (table 7.2). Breeding systems range from obligate outcrossers (e.g., *Brassica campestris*, *Coffea canephora*) to highly selfing plants (e.g., *Triticum* spp., *Arachis hypogaea*). Some are animal-pollinated, others wind-pollinated. Some are capable of vegetative reproduction as well.

This review updates and expands work that I previously published in a collaborative study on the world's thirteen most widely grown crops based on 1997 data (Ellstrand et al. 1999). The growing body of evidence for spontaneous hybridization is reflected in the many new studies that have emerged

Table 7.1 The World's Most Important Food Crops

Rank	Crop	Area Planted[a] (10^6 Hectares)	Scientific Name	Family
1	Wheat	213.8	*Triticum aestivum*[b]	Poaceae
			T. turgidum durum[b]	Poaceae
2	Rice	154.1	*Oryza sativa*	Poaceae
			O. glaberrima	Poaceae
3	Maize (sweet and field corn)	139.7	*Zea mays mays*	Poaceae
4	Soybean	74.1	*Glycine max*	Fabaceae
5	Barley	53.6	*Hordeum vulgare*	Poaceae
6	Sorghum	42.0	*Sorghum bicolor bicolor*	Poaceae
7	Millet	36.3	*Eleusine coracana coracana*[b]	Poaceae
			Pennisetum glaucum[b]	Poaceae
8	Cottonseed	31.6	*Gossypium hirsutum*[b]	Malvaceae
			G. barbadense[b]	Malvaceae
9	Rapeseed (canola)[c]	25.7	*Brassica napus*	Brassicaceae
			B. campestris	Brassicaceae
10	Beans, dry, green, and string	25.0	*Phaseolus vulgaris*[b]	Fabaceae
11	Groundnut (peanut)	24.3	*Arachis hypogaea*	Fabaceae
12	Sunflower seed	21.1	*Helianthus annuus*	Asteraceae
13	Potato	19.9	*Solanum tuberosum*[b]	Solanaceae
14	Sugarcane	19.2	*Saccharum officinarum*[b]	Poaceae
15	Cassava	17.0	*Manihot esculenta*	Euphorbiaceae
16	Oats	12.8	*Avena sativa*	Poaceae
17	Coconut	11.6	*Cocos nucifera*	Arecaceae
18	Coffee	10.9	*Coffea arabica*[b]	Rubiaceae
			C. canephora[b]	Rubiaceae
19	Chickpea	10.0	*Cicer arietinum*	Fabaceae
20	Cowpea	9.9	*Vigna unguiculata unguiculata*	Fabaceae
21	Rye	9.7	*Secale cereale*	Poaceae
22	Oil palm fruit	9.6	*Elaeis guineensis*	Arecaceae
23	Sweet potato	9.1	*Ipomoea batatas*	Convolvulaceae
24	Olive	8.0	*Olea europaea europaea*	Oleaceae
25	Grape	7.7	*Vitis vinifera*[b]	Vitaceae
TOTAL AREA		996.7		

[a]Area of production for 2000 from the FAOSTAT website, http://apps.Fao.org/ (6/6/2002).
[b]Other taxa account for a small portion of world production of this crop.
[c]Does not include the same species grown as vegetable or fodder crops.

Table 7.2 *The World's Most Important Food Crops: Reproductive Biology*

Crop	Scientific Name	Primary Pollen Vector(s)	Typical Breeding System[a]	Capable of Vegetative Spread?	Crop Requires Flowering?
Wheat	*Triticum aestivum*[b]	Self, wind	Selfing	No	Yes
	T. turgidum durum[b]	Self, wind	Selfing	No	Yes
Rice	*Oryza sativa*	Self, wind	Selfing	No	Yes
	O. glaberrima	Self, wind	Selfing	No	Yes
Maize (sweet and field corn)	*Zea mays mays*	Wind	Outcrossing	No	Yes
Soybean	*Glycine max*	Self, insects	Selfing	No	Yes
Barley	*Hordeum vulgare*	Self, wind	Selfing	No	Yes
Sorghum	*Sorghum bicolor bicolor*	Self, wind	Selfing	No[c]	Yes
Millet	*Eleusine coracana coracana*[b]	Self, wind	Selfing	No	Yes
	Pennisetum glaucum[b]	Wind	Outcrossing	No	Yes
Cottonseed	*Gossypium hirsutum*[b]	Self, insects	Selfing	No	Yes
	G. barbadense[b]	Self, insects	Selfing	No	Yes
Rapeseed (canola)	*Brassica napus*	Self, insects	Selfing	No	Yes
	B. campestris	Insects	Outcrossing	No	Yes
Beans, dry, green, and string	*Phaseolus vulgaris*[b]	Self, insects	Selfing	No	Yes
Groundnut (peanut)	*Arachis hypogaea*	Self	Selfing	No	Yes
Sunflower seed	*Helianthus annuus*	Insects, self	Outcrossing	No	Yes
Potato	*Solanum tuberosum*[b]	Insects	Selfing	Yes	No
Sugarcane	*Saccharum officinarum*[b]	Wind	Outcrossing	Yes	No
Cassava	*Manihot esculenta*	Insects, self	Outcrossing, apomixis	Yes	No
Oats	*Avena sativa*	Self, wind	Selfing	No	Yes
Coconut	*Cocos nucifera*	Wind, self, insects	Varies	No	Yes
Coffee	*Coffea arabica*[b]	Insects, self	Selfing	No	Yes
	C. canephora[b]	Insects	Outcrossing	No	Yes
Chickpea	*Cicer arietinum*	Self, insects	Selfing	No	Yes
Cowpea	*Vigna unguiculata unguiculata*	Self, insects	Selfing	No	Yes
Rye	*Secale cereale*	Wind	Outcrossing	No	Yes
Oil palm fruit	*Elaeis guineensis*	Insects	Outcrossing	No	Yes
Sweet potato	*Ipomoea batatas*	Insects	Outcrossing	Yes	No
Olive	*Olea europaea europaea*	Insects	Outcrossing	No	Yes
Grape	*Vitis vinifera*[b]	Insects	Outcrossing	No	Yes

[a]From Frankel and Galun 1977, Fryxell 1957.
[b]Other taxa account for a small portion of the world production of this crop.
[c]But certain compatible wild relatives are capable of vegetative spread.

since that paper was written (e.g., Abe et al. 1999, Alexander et al. 2001, Guad-agnuolo et al. 2001a, 2001b, Hansen et al. 2001).

Finding information about natural hybridization between crops and their wild relatives is a challenge. Because the topic is the concern of many different kinds of plant biologists, the data are spread thinly throughout the literature of many diverse sciences, ranging from evolution and systematics to economic botany, agronomy, and horticulture. While some data are readily accessible from journal articles, just as much seems to be published in less accessible literature such as books, reports, and symposium proceedings. After compiling my conclusions for each crop, I consulted one or more experts (see the acknowledgments) to preview my conclusions, to suggest other experts, and to identify gaps in my treatment. However, I take final responsibility for the interpretation of the data.

The crops listed below start with the most widely planted crop, wheat, and are then presented in decreasing order of importance. Each crop review starts with a brief general description of the species to be covered, with the scientific nomenclature generally following that of Smartt and Simmonds (1995). I list some of the crop's uses, its region of origin, and its current areas of cultivation, relying mostly on Hancock (1992), Simpson and Ogorzaly (2001), and Smartt and Simmonds (1995) for details.

Then I enumerate those wild relatives likely to naturally hybridize with the crop and their geographic distribution. I present the evidence for natural hybridization and introgression for each crop. Data regarding the fitness of hybrids or their descendants relative to the pure wild type under field conditions, if available, are presented because those data indicate the likelihood of penetration and persistence of crop alleles in natural populations (see chap. 4). High-priority research gaps, if appropriate, are identified for most crops. I also report the known or suspected impacts of hybridization, such as the evolution of increased weediness, the enhanced risk of extinction by hybridization, or the evolution of a new species. The chapter concludes with a summary examining the trends that emerge from the review. I abstract my conclusions in table 7.3.

I have attempted to present the contemporary "state of the art" for each crop. But for many crops it may be impossible to be complete, given the rapidly growing interest in the field, fueled by concerns about the escape of crop transgenes into natural populations (e.g., Hancock et al. 1996, Snow and Moran-Palma 1997, Wolfenbarger and Phifer 2000). By the time you read this, at least a half-dozen new cases of crop-to-wild hybridization and introgression will no doubt be waiting to find their place in the next edition.

#1. Wheat

In terms of area planted and tons harvested, wheat is the world's most important crop. The allopolyploid annual grasses *Triticum aestivum* L. ssp. *aestivum* (bread wheat) and *T. turgidum* L. ssp. *durum* (Desf.) Husn. (durum wheat) are the most important of the six wheat species (following the nomenclature of Van Slageren 1994). *T. aestivum* is an allohexaploid species (2n = 6× = 42 chromosomes); *T. turgidum* ssp. *durum* is an allotetraploid (2n = 4× = 28 chromosomes). Originating in southwestern Asia, both species have been introduced to all the inhabited continents and are of particular importance in Asia, Europe, and North America. China produces more of the world's wheat (ca. 26 percent) than any other nation. The primary product of both species is flour. Flour from *T. aestivum* is used for most of humankind's breads and pastries. Wheat-based pasta products, from angel hair to ziti, are made of flour or semolina from durum wheat. A few other *Triticum* taxa are cultivated (Zohary and Hopf 1993), but they account for only a tiny fraction of the world's wheat production and will not be considered here.

Wild relatives of wheat include the allied genus *Aegilops* (goatgrass) and four wild taxa of *Triticum*. Most of these are native to southwestern Asia (Feldman et al. 1995). Several have extended their ranges to other continents within the past few thousand years with unintentional human assistance. In fact, certain allotetraploid *Aegilops* species have become serious, or even noxious, weeds (e.g., Anderson 1993).

Plants appearing to be hybrids between cultivated wheat and its wild relatives occur frequently on the margins of wheat fields when the wild *T. turgidum* subspecies or certain species of *Aegilops* are present (e.g., Kimber and Feldman 1987, Ladizinsky 1992, Popova 1923, Van Slageren 1994). Most reports are from the Mediterranean basin and the Middle East, but some are from elsewhere in Africa, Europe, Asia, and even North America. At least eleven different *Aegilops* species apparently hybridize naturally with bread wheat: *Ae. biuncialis, Ae. crassa, Ae. cylindrica, Ae. geniculata, Ae. juvenalis, Ae. neglecta, Ae. speltoides* var. *ligustica, Ae. tauschii, Ae. triuncialis, Ae. umbellulata,* and *Ae. ventricosa* (reviewed by Van Slageren 1994). At least four *Aegilops* species *(Ae. columnaris, Ae. geniculata, Ae. neglecta,* and *Ae. ventricosa)* are thought to spontaneously hybridize with durum (reviewed by Van Slageren 1994), but the list is not likely to be exhaustive (M. W. Van Slageren, personal communication). In addition to their morphological intermediacy, the fact that these plants are fully or partially sterile supports their presumed hybridity.

Although breeders have produced fertile hybrids between wheat and its wild relatives (Feldman et al. 1995, Jiang et al. 1994), all natural hybrids are typically highly sterile, although seeds may occasionally be found (Popova 1923, Van Slageren 1994). For example, viable seed set on hybrids of *Triticum aestivum* × *Aegilops cylindrica* (jointed goatgrass) is very limited (Zemetra et al. 1998, Snyder et al. 2000, Guadagnuolo et al. 2001a). This hybrid sterility may explain why hybridization generally appears to be restricted to F_1s with little evidence for subsequent introgression. Nonetheless, second-generation backcrosses of bread wheat and jointed goatgrass that were generated by hand in the greenhouse showed a substantial recovery in female fitness relative to prior hybrid generations (Zemetra et al. 1998; see also Guadagnuolo et al. 2001a). Furthermore, both first- and second-generation backcrosses of bread wheat and jointed goatgrass have been obtained from field-grown plants, and these too showed a restoration of fertility relative to the hybrids (Mallory-Smith et al. 1999). If a few individuals with hybrid history were able to persist and reproduce, a limited amount of introgression could still occur (Arnold et al. 1999).

Indeed, introgression from cultivated durum into wild emmer, *T. dicoccoides,* has been implicated in the evolution of a distinct race of wild emmer in the Upper Jordan Valley. At present, wild emmer in the Upper Jordan Valley has a number of morphological, physiological, and isozyme traits that are common in durum and rare or absent in other wild emmer populations (Blumler 1998). The Upper Jordan Valley has a long history of durum cultivation, thousands of years of opportunity for hybridization and introgression.

RAPD and microsatellite progeny analysis of *Ae. cylindrica,* experimentally planted within a field of *T. aestivum* in Switzerland, detected spontaneous hybridization that varied with the source population of the goatgrass (Guadagnuolo et al. 2001a). One percent of the seeds were hybrids for goatgrass grown from seed collected from two long-established Swiss populations. Seven percent of the seeds were hybrids for goatgrass collected from a recently established Swiss population.

A descriptive study was conducted to determine whether any wheat alleles had introgressed into populations of more distant relatives, *Elymus caninus* and *Hordeum marinum,* growing in the vicinity of English and Austrian wheat fields (Guadagnuolo et al. 2001b). Seeds from these wild plants were germinated and grown for analysis of their morphology, isozymes, RAPDs, and microsatellites. "No evidence of introgression of wheat traits into *E. caninus* was observed. However, one individual of *H. marinum* which had the typical morphology of this species" had seven DNA markers specific to wheat; the combination of these characters is that expected for hybrid ancestry (Guadagnuolo et al. 2001b).

Gene flow between crops and their wild relatives is generally considered to occur with the wild plant as the seed parent (Ladizinsky 1985). However, two herbicide-resistant hybrids of *T. aestivum* × *Ae. cylindrica* appeared in seed collected from herbicide-resistant wheat plants grown as a field trial in the United States (Seefeldt et al. 1998). These hybrids were the result of natural pollination of the crop parents by nearby stands of wild goatgrass. In this case the domesticated plant is the seed parent. The hybrids were taller and more robust than their domesticated parent but were largely seed-sterile, producing an average of 3.5 seeds per plant.

Clearly, cultivated wheat can and does occasionally hybridize naturally with certain wild relatives. The extent of introgression is presumed to be limited to highly sterile hybrids. A thorough population-genetic analysis of some of these hybrid swarms would reveal whether that is indeed the case.

#2. Rice

In terms of the number of people fed, rice is the world's most important crop. The world's rice production is based on two annual grass species. *Oryza sativa* L., originally from southeastern Asia, is widely planted throughout the humid tropics and subtropics of the world. It is gradually replacing the other cultivated rice species, *O. glaberrima* Steud., a native of Africa grown mostly on that continent. Rice is consumed mostly as a boiled grain. It is a particularly important crop in Asia; China produces more of the world's rice (ca. 34 percent) than any other nation.

Both cultivated rice species are interfertile with certain close wild relatives. For *O. sativa,* they are weedy conspecifics (*O. sativa* f. *spontanea*) as well as *O. nivara* and *O. rufipogon.* In the case of *O. glaberrima,* the compatible wild relatives are *O. barthii* and *O. longistamina.* Unintended human assistance has helped the wild relatives of rice expand their ranges to all of the inhabited continents, often becoming serious weeds of both cultivated rice and other crops (Holm et al. 1997). Hand crosses between the crop species and their wild counterparts are easily accomplished (e.g., Chu and Oka 1970, Morishima et al. 1992). The fitness of hybrid progeny from crosses between wild and cultivated rice is generally high, but hybrids from certain crosses show reduced fertility (e.g., Chu et al. 1969). In contrast, when the two cultivated species are intercrossed, the hybrids are highly sterile (Chang 1995).

There is ample evidence that both *O. sativa* and *O. glaberrima* naturally hybridize with their wild relatives. Plants appearing to be hybrids or hybrid derivatives of both cultivated rice species and their wild relatives occur frequently in and near rice fields when wild taxa are present, particularly in Asia and Africa but also in North America and South America. These morphologically intermediate plants usually appear as hybrid swarms. In the case of *O. sativa* the intermediates may sometimes be relatively uniform and behave as stabilized races. In numerous studies genetic analysis of the wild intermediates has been conducted using isozymes, RAPDs, cytogenetic analysis, and progeny segregation studies of morphological and physiological traits (e.g., Chu and Oka 1970, Majumder et al. 1997, Oka 1988, Oka and Chang 1961, Suh et al. 1997, Tang and Moroshima 1988). Those analyses have shown repeatedly that intermediate plants have a combination of alleles specific to both the pure crop and the pure wild taxon, supporting the hypothesis of their hybrid ancestry. For example, Second (1982) genetically analyzed some of the weedy populations of *O. barthii* in Africa that have morphological similarities to *O. glaberrima*. He found that the weeds contain allozyme alleles typical of *O. glaberrima* but unknown in other wild populations of *O. barthii*.

Natural rates of hybridization between wild and cultivated rice can be substantial. In an experimental study, seed was collected from the weed red rice *(O. sativa f. spontanea)*, growing intermixed with cultivated *O. sativa* in experimental plots in Louisiana. Morphological and allozyme analysis of the progeny revealed rates of natural hybridization ranging from 1 percent to 52 percent (Langevin et al. 1990). Hybridization rates varied depending on which rice cultivar was the pollen parent; those cultivars whose flowering season overlapped the most with that of the weed hybridized with it the most. The hybrids demonstrated heterosis; they were generally taller and produced more "tillers" (as vegetative offshoots in the grass family are called) than either parental type (Langevin et al. 1990).

A notable example of crop-to-wild gene flow involving rice occurred during the last century. To facilitate weeding, a purple-pigmented rice cultivar was planted in India to distinguish its seedlings from the green seedlings of the weedy *O. sativa* f. *spontanea*. The strategy was thwarted by introgression. After a few seasons of gene flow and selection, the weed populations had accumulated the pigmentation allele at a high frequency (Dave 1943, Oka and Chang 1959, Harlan et al. 1973, Parker and Dean 1976).

Natural hybridization with cultivated rice has been implicated in the near extinction of the endemic Taiwanese taxon *O. rufipogon* ssp. *formosana* (Kiang et al. 1979, Oka 1992). Collections of this wild rice over the last century show a progressive shift toward characters of the cultivated species and a coincidental decrease in fertility of seed and pollen. By 1979, naturally occurring populations of wild rice in Taiwan were on the verge of extinction. Indeed, throughout Asia typical specimens of other subspecies of *O. rufipogon* and the wild *O. nivara* are now rarely found because of extensive spontaneous hybridization with the crop (Chang 1995).

#3. Maize (Corn)

The annual grass *Zea mays* ssp. *mays* L. produces maize. An amazingly versatile crop, it is grown as a vegetable for human consumption, for livestock feed, and for popcorn, as well as being processed for flour, meal, oil, sugar, syrup, and alcohol. Native to Mexico, it has become important on a global scale. Still, production remains largely confined to the New World. North America and Central America alone produce almost half of the world's maize. The United States leads the world in maize production, accounting for about 40 percent of the crop (interestingly, China is second).

The closest wild relatives of maize are the various teosintes (Sánchez González and Ruiz Corral 1997). Teosintes comprise the remaining taxa in the genus *Zea*. Some are wild subspecies of *Z. mays;* others are distinct species. Teosintes are distributed from the state of Chihuahua in northern Mexico south to Nicaragua (Sánchez González and Ruiz Corral 1997). Most teosintes can be readily hand-crossed with maize to yield fertile hybrids (Goodman 1995); crosses with *Z. mays* ssp. *mexicana* are more difficult (Kermicle 1997). Maize × *Z. perennis* hybrids are typically highly sterile (Doebley 1990). Maize can also be crossed, with great difficulty, with members of the genus *Tripsacum* (Goodman 1995). While they are occasionally perceived as weeds in certain locales, an increasing number of teosinte populations are at risk of extinction. In fact, hybridization with maize may have played a role in the extinction of the populations that were maize's progenitors (Small 1984).

For years the question of whether maize alleles had introgressed into teosinte populations was controversial. Plants appearing to be hybrids between maize and teosinte often occur spontaneously in and near Mexican maize fields where teosinte, particularly *Z. mays* ssp. *mexicana,* is abundant (Wilkes 1977). Allozyme analysis of accessions of the teosintes *Z. luxurians, Z. diploperennis,* and *Z. mays* ssp. *mexicana* revealed that alleles that are otherwise maize-specific occurred at extremely low frequencies, suggesting a very low level of introgression from maize into these teosinte taxa (Doebley 1990). DNA sequences from the nuclear ribosomal internal transcribed space region of all *Zea* taxa show a similar pattern (Buckler and Holtsford 1996). For example, one allele from a *Z. mays* ssp. *parviglumis* plant clustered with those from the crop. Such anomalous groupings may be the result of past hybridization, or alternatively they may be the continued segregation of ancestral polymorphisms (symplesiomorphy; see chap. 5). In contrast, cytogenetic analyses "offer no evidence of . . . maize-teosinte introgression in either direction" (Kato 1997). Allozyme analyses of accessions or bulk seed collections of teosinte growing together with maize showed no evidence of introgression from the crop into the wild taxa (Doebley 1990). Such comparisons are not available for *Z. mays* ssp. *parviglumis,* for which no crop-specific allozyme alleles are apparently available.

The population-genetic work by Blancas (2001) offers the most compelling evidence for the introgression of maize alleles into populations of teosinte. She collected from maize and teosinte *(Z. mays* ssp. *mexicana)* populations that were growing intermingled in central Mexico, carefully segregating the morphologically intermediate individuals for separate analysis. She also collected from allopatric populations of both taxa for comparison. She genetically analyzed the collections for eighteen allozyme loci. When genetic distances were calculated, she found that the pairs of locally sympatric populations of teosinte and maize were more similar to one another than to any of the other populations, including members of the same taxon. Morphologically intermediate populations had intermediate allele frequencies at some loci but at others had frequencies that transgressed those of the morphologically "pure" populations of local parental types. These patterns are compatible with the hypothesis of hybridization and introgression of maize alleles into teosinte. They also suggest that the morphologically intermediate populations may be evolving as a third lineage, diverging from the parental taxa.

No evidence suggests that natural hybridization occurs between maize and *Z. perennis* and *Tripsacum* (but see Talbert et al. 1990 for a hybrid derivative of teosinte and a *Tripsacum* species).

#4. Soybean

Soybeans are the seeds of the annual legume *Glycine max* (L.) Merr. Native to East Asia, this species became a crop of worldwide importance only during the twentieth century. The crop is particularly important in North America, South America, and Asia. The United States accounts for about 38 percent of the world's soybean production. Although the seeds can be cooked and consumed directly, they are more often processed to create a variety of products, including ones for human consumption (e.g., tofu, miso, soy sauce, synthetic meat substitutes, soybean oil), for livestock consumption, and for industrial use (e.g., soy-based ink). Its only compatible wild relatives are *G. soja* and *G. gracilis* (see below). Their distribution is restricted to peninsular Korea, Taiwan, Japan, northeastern China, and the adjacent part of Siberian Russia (Singh and Hymowitz 1999).

Wild plants morphologically intermediate to *G. max* and *G. soja* often occur spontaneously near Asian soybean fields when *G. soja* is present (Abe et al. 1999). The apparently stabilized hybrid taxon *G. gracilis* (= *G. max* forma *gracilis*) is intermediate to and interfertile with both *G. max* and *G. soja*. The three are sometimes considered the same species (Singh and Hymowitz 1989). *G. gracilis* accessions in germplasm collections have been genetically characterized with morphological, chemical, and RFLP markers; all studies have shown that the accessions have a mixture of alleles diagnostic for both putative parents (e.g., Broich and Palmer 1981, Keim et al. 1989). However, population-level analysis has not yet been conducted to determine the extent to which crop alleles have moved into natural populations. It is not yet clear whether *G. gracilis* represents a stabilized hybrid-derived taxon or a subset of the variation found within hybrid swarms between *G. max* and *G. soja,* as self-fertilizing accessions maintained in germplasm collections.

#5. Barley

The annual grass *Hordeum vulgare* L. produces barley. Native to southwestern Asia, this crop is now grown in the world's cool temperate regions, particularly North America, Europe, and Asia. Germany leads in barley production, accounting for more than 10 percent of the world's crop. The grain is used directly for human and livestock consumption but is also used in the brewing of beer and whiskey.

H. vulgare crosses easily to form fertile hybrids with its wild, weedy progenitor *H. spontaneum* (= *H. vulgare* ssp. *spontaneum*), whose distribution ranges from the eastern Mediterranean to Iran and West Central Asia (Harlan 1995). Barley can be crossed with *Hordeum* species other than *H. spontaneum,* but the resulting hybrids are typically highly sterile.

Wild plants morphologically intermediate to cultivated barley and *H. spontaneum* often "occur where the two species are found together" (Harlan 1995) "in less intensively cultivated fields or areas adjacent to cultivated fields" in the Middle East (von Bothmer et al. 1991). Despite morphological evidence for hybridization, I am not aware of any genetic analysis of the spontaneous intermediates.

#6. Sorghum

The annual grass *Sorghum bicolor* ssp. *bicolor* (L.) Moench is native to Africa and now cultivated in hot, dry regions of the world, particularly Africa, Asia, and North America. Both its seed and its vegetative parts are used for food for both humans and livestock. It is the major human food grain in parts of India and Africa. Human-consumed sorghum products include syrup processed from the plant's stems and flour processed from its seeds. The United States is the world's number one sorghum-producing country, producing roughly 22 percent of the crop.

The crop's closest compatible wild relatives are the wild subspecies *S. bicolor* ssp. *drummondii* and *S. bicolor* ssp. *verticilliflorum,* which are also native to Africa (Doggett and Prasada Rao 1995). The crop is also interfertile with *S. propinquum* of southeastern Asia and *S. halepense,* which is native to southwestern Asia and adjacent North Africa. Both of these species are perennials, capable of vegetative spread and persistence with well-developed rhizomes. Also, both have spread to other continents as weeds. In fact, *S. halepense,* johnsongrass, is considered to be one of the world's worst weeds (Holm et al. 1977).

Plants that are morphologically intermediate between cultivated sorghum and the wild relatives mentioned in the last paragraph occur frequently in and near sorghum fields when wild taxa are present, in both the Old World and the New World (Baker 1972, Doggett and Majisu 1968). When these wild plants are genetically analyzed, they are found to hold crop-specific alleles. Analyses of progeny segregation, allozymes, and RFLPs all reveal crop-specific alleles in populations of wild *S. bicolor* when it co-occurs with the crop in Africa, suggesting that intraspecific hybridization and introgression are common (Aldrich and Doebley 1992, Aldrich et al. 1992, Doggett and Majisu 1968).

Even though the crop is a diploid species (2n = 20 chromosomes) and the weed *S. halepense* is an allotetraploid (2n = 4× = 40 chromosomes), natural hybridization between them can readily occur. Allozyme progeny analysis of *S. halepense,* experimentally planted around cultivated sorghum fields in California, detected spontaneous hybridization as far as one hundred meters from the crop with "rates as high as 2%" at that distance (Arriola and Ellstrand 1996). The resulting hybrids were compared with nonhybrid johnsongrass under field conditions. Hybrids showed no significant difference from the weed for a number of fitness correlates, including seed production, pollen viability, and tiller production (Arriola and Ellstrand 1997). Despite the differences in chromosome numbers between the two species, relatively high fertility in the hybrids is not as surprising as might be first thought because chromosome counts of johnsongrass × sorghum hybrids report predominantly tetraploids (reviewed in Warwick and Black 1983). Also, it should be noted that the opportunity for perenniality and vegetative spread in johnsongrass × sorghum hybrids might foster their persistence, regardless of their sexual fertility (cf. Ellstrand et al. 1996).

Hybridization involving this crop has been implicated in the evolution of weediness in wild *Sorghum. S. almum,* which is both a weed and a forage grass, is apparently a stabilized hybrid of crop sorghum and *S. propinquum.* Likewise, introgression from crop sorghum has been implicated in the evolution of enhanced weediness in *S. halepense* in North America, South Asia and elsewhere (Holm et al. 1977). Genetic analysis of *S. almum* and *S. halepense* using RFLPs supports their proposed hybrid ancestries (Paterson et al. 1995).

#7. Millet

The word "millet" is used to refer to grain produced by several unrelated tropical annual grasses. As a group, significant production of millet is restricted to Africa and Asia, particularly India. The two most important species are finger millet, *Eleusine coracana* ssp. *coracana* (L.) Gaertn., and pearl millet, *Pennisetum glaucum* (L.) R. Br. (= *P. americanum* = *P. typhoides*) (Hancock 1992).

Finger Millet

Native to Africa, finger millet is grown primarily in eastern equatorial Africa and southern India. The seeds are typically processed into flour for bread or eaten as porridge.

The only close relative that readily crosses with finger millet is its wild counterpart, *E. coracana* ssp. *africana* (de Wet 1995a), which is found primarily in Africa. Plants with a range of morphologies between finger millet and *E. coracana* ssp. *africana* are often found along roadsides as well as in, and at the edges of, finger millet fields where the two taxa co-occur in Africa (Mehra 1962, Phillips 1972). These spontaneous intermediates sometimes behave as noxious weeds (de Wet 1995a). Despite morphological evidence for hybridization, I am not aware of any genetic analysis of the spontaneous intermediates.

Finger millet is also closely related to one of the world's worst weeds, *E. indica* (Holm et al. 1977), but strong cross-incompatibility barriers isolate the two (de Wet 1995a). Natural hybrids between finger millet and *E. indica* or between finger millet and other wild *Eleusine* have not been reported.

Pearl Millet

Native to Africa, pearl millet is grown primarily in Africa, India, and Pakistan. The seeds are typically eaten as a boiled grain, processed into flour, or malted.

Pearl millet is closely related to and cross-compatible with two wild species that are native to Africa, *P. violaceum* (= *P. glaucum* ssp. *monodii*) and *P. sieberanum* (= *P. glaucum* ssp. *stenostachyum*) (de Wet 1995b). *P. violaceum* is a wild species that occurs independently of agriculture. *P. sieberanum* (shibra) is a wild, weedy crop mimic that occurs only in cultivated fields.

Plants appearing to be hybrids between pearl millet and *P. violaceum* or between pearl millet and *P. sieberanum* often occur in or near pearl millet fields in West Africa and northern Namibia when the wild taxa are present (Brunken et al. 1977). Allozyme analysis of *P. violaceum* accessions revealed that populations intermingled with the crop were genetically more similar to the crop than were populations growing a substantial distance from the crop (Tostain 1992), supporting the hypothesis of crop-to-wild introgression in areas of close contact.

Hybridization between pearl millet and *P. violaceum* may have been involved in the evolution of *P. sieberanum,* which is morphologically intermediate to the two species. "In Africa, shibras are found throughout much of the area of pearl millet cultivation" (Brunken et al. 1977), where, as "obligate weeds of cultivation," they "do not persist for more than one generation after cultivated fields have been abandoned" (de Wet 1995b).

Natural hybridization between wild and cultivated *Pennisetum* can occur at a substantial rate. Genetic markers (allozyme and morphological) were used to measure hybridization between naturally adjacent wild and cultivated populations of millet in Niger (Marchais 1994). The crop sired about 35 percent of the progeny of both *P. violaceum* and *P. sieberanum.* Also, a plot of interplanted wild and cultivated millet was genetically structured so that allozyme progeny analysis would identify hybrids (Renno et al. 1997). In this experiment, also conducted in Niger, the crop sired 8 percent of the progeny of *P. violaceum* and 39 percent of the progeny of *P. sieberanum.*

Pearl millet is also closely related to one of the world's worst weeds, *P. purpureum* (Holm et al. 1977), but the two are isolated by strong cross-incompatibility barriers (de Wet 1995b). Natural hybrids between pearl millet and *P. purpureum* or between pearl millet and other wild *Pennisetum* have not been reported.

#8. Cottonseed

Because cotton is grown primarily for fiber production, it isn't popularly thought of as a food crop. But humans and animals consume lots of cotton products. Cottonseed is one of the world's most important sources of vegetable oil. Additionally, cottonseed is processed into meal. Interestingly, the fiber is sometimes processed for human consumption into methylcellulose, an additive in numerous processed foods, including ice cream. Four perennial shrub species are typically grown as annuals to produce cotton. The most important species are *Gossypium hirsutum* L. and *G. barbadense* L., both of which originated in the New World (Mesoamerica and northwestern South America, respectively). These two species dominate worldwide cotton production, having largely displaced those cotton species domesticated in the Old World. China produces more of the world's cottonseed (ca. 22 percent) than any other nation, and more than half of the world's production comes from Asia. Two other *Gossypium* species are cultivated but account for only a tiny fraction of the world's cotton production and will not be considered here.

Gossypium comprises about fifty species. Wild cottons most closely related to the two most important cultivated species are wild or feral forms of *G. hirsutum* and *G. barbadense,* which occur in Central and South America, the Caribbean, and some Pacific Island groups. Other closely related wild taxa are *G. darwinii* of the Galapagos Islands, *G. mustelinum* of northeastern Brazil, and *G. tomentosum* of the Hawaiian Islands (Wendel 1995).

Allozyme and DNA analyses have demonstrated that limited interspecific introgression has occurred from the cultivated cotton species into these wild relatives. *G. barbadense*–specific alleles were found in low frequency in wild or feral populations of *G. hirsutum* that are sympatric with the crop in Mesoamerica and the Caribbean. In the same regions *G. hirsutum*–specific alleles occur in wild *G. barbadense* populations that are sympatric with cultivated *G. hirsutum* (Brubaker et al. 1993, Brubaker and Wendel 1994, Wendel et al. 1992). Population-level analysis should be conducted to determine the extent to which crop alleles have moved into natural populations. Intraspecific gene flow from these crops to their wild forms has apparently not yet been investigated.

Fryxell (1979) suggested, on the basis of morphology, that both *G. darwinii* of the Galapagos Islands and *G. tomentosum* of the Hawaiian Islands were at risk of extinction as a result of hybridization with *G. hirsutum*. The presence of *G. hirsutum*–specific allozyme alleles in wild populations has confirmed that *G. darwinii* has experienced substantial introgression from the crop (Wendel and Percy 1990). Interestingly, that gene flow appears to have come not directly from *G. hirsutum* but from hybridization with introduced *G. barbadense* that had a history of introgression from *G. hirsutum*. This situation represents, to my knowledge, the only known case of natural "bridge" introgression from a crop to a wild relative. Attempts to determine whether interspecific hybridization between *G. hirsutum* and *G. tomentosum* occurs have been hampered by the lack of species-specific markers (DeJoode and Wendel 1992). Allozyme analysis has revealed limited introgression of *G. hirsutum* alleles into the Brazilian endemic *G. mustelinum* (Wendel et al. 1994).

#9. Rapeseed (Canola)

Rapeseed is the oil crop of two annual herbs, *Brassica napus* L. and *B. campestris* L. (also known as *B. rapa*). Other cultivars of *B. napus* and *B. campestris* are grown as forages and vegetables, notably swedes (rutabagas) in the case of *B. napus* and Chinese cabbage and turnip in the case of *B. campestris*. The diploid (2n = 20 chromosomes) *B. campestris* is the less important of the two rapeseed species. It appears to have been domesticated both in the Mediterranean region of Europe and in the eastern Afghanistan/western Pakistan region of Asia (McNaughton 1995b). Although its place of origin is unknown, *B. napus* is known to be an allotetraploid (2n = 4× = 38 chromosomes), a natural hybrid derivative of *B. campestris* and *B. oleracea* (the latter species accounts for many cole crops from cabbage to broccoli) (McNaughton 1995a). Rapeseed production has risen dramatically in the last few decades. It is a crop of cool, temperate regions, primarily in Asia, Europe, and North America. China leads the world in production, producing roughly 23 percent of the crop.

The closest wild relative to both rapeseed species is *B. campestris* ssp. *eu-campestris,* a serious and widespread weed of Eurasia, where it is native, as well as North America, Australia, New Zealand, and temperate parts of South America (Holm et al. 1997). "It is uncertain whether or not *B. napus* exists in truly wild form" (McNaughton 1995a), but it has naturalized in numerous locations, including the British Isles (Stace 1991), France (Pessel et al. 2001), Scandinavia (Mossberg et al. 1992), and North America (e.g., Hickman 1993). Many other species are potentially natural mates with oilseed rape. Most of these are native to the Old World and have spread globally as serious weeds (Holm et al. 1997). Some are wild *Brassicas;* others are members of allied genera such as *Hirschfeldia* and *Raphanus.*

Brassica campestris *Oilseed Rape*

Spontaneous hybridization between a *B. campestris* vegetable cultivar and wild *B. campestris* ssp. *eu-campestris* has been experimentally measured (Manasse 1992). Stands of a cultivar homozygous for a dominant anthocyanin (pigmentation) marker allele were planted in Washington State. At varying distances around these were planted individuals of wild *B. campestris* ssp. *eu-campestris* homozygous for the recessive allele. Progeny testing from the wild plants revealed that this vegetable cultivar of *B. campestris* readily hybridizes with its weedy conspecific under field conditions. I am not aware of any descriptive or experimental study addressing whether oilseed *B. campestris* can spontaneously hybridize with wild *B. campestris*. However, hybridization rates involving *B. campestris* oilseed cultivars probably will be similar to those for the vegetable cultivar.

Brassica napus *Oilseed Rape*

The most likely recipients of gene flow from cultivated *B. napus* ought to be naturalized, weedy populations of the same species. Although reproductive barriers that would prevent spontaneous hybridization between weedy and cultivated *B. napus* are unlikely, spontaneous hybridization rates between the two forms have apparently not been measured.

In contrast, there is abundant evidence that cultivated *B. napus* spontaneously hybridizes with wild *B. campestris* ssp. *eu-campestris* and that introgression occurs after hybridization as well. Their morphologically intermediate, partially sterile hybrid *(B. × harmsiana)* occurs sporadically in the British Isles and elsewhere in Europe in or near crops of *B. napus* growing adjacent to or intermixed with wild *B. campestris* (Harberd 1975). Likewise, Jørgensen and Andersen (1994) found *B. napus*–specific allozyme alleles in two wild *B. campestris*–like plants in Denmark—evidence of past hybridization and introgression. Jørgensen et al. (1998) analyzed a population of *B. campestris* growing in a Danish field of volunteer oilseed rape and identified a number of likely hybrids and backcrosses based on a combination of morphology, chromosome counts, isozymes, and RAPDs. Recently, Hansen et al. (2001) used species-specific AFLPs to analyze plants collected from a Danish population of wild *B. campestris* mixed with feral *B. napus* that had persisted eleven years after intentional cultivation. Of the 102 plants collected from a plot three meters square, 1 plant had the genetic profile of a first-generation hybrid and 44 were advanced-generation hybrids, having markers specific to both parental species. The remarkable level of introgression in this population is less of a surprise than might be predicted from differences in chromosome number in light of the findings of recent experiments measuring spontaneous hybridization rates between the two species and experiments measuring the fitness of their hybrids and hybrid derivatives.

Field-based experiments in Denmark by Jørgensen and colleagues have used a variety of genetically based markers to measure spontaneous hybridization rates between *B. napus* and wild *B. campestris*. *B. campestris* growing within stands of the crop produced anywhere from 9 percent to 93 percent hybrid progeny, depending on the experimental design (Jørgensen et al. 1996). The same research group later planted stands of *B. napus* oilseed rape transgenic for herbicide resistance intermixed with wild *B. campestris*. Both species produced spontaneous transgenic interspecific hybrids; "the frequency of hybrids was 3% and 0.3% with the weedy species and oilseed rape as female, respectively" (Jørgensen et al. 1998).

The transgenic hybrids were grown in plots with wild *B. campestris* to assess the potential for backcrossing to the wild parent. Eight hundred sixty-five herbicide-resistant progeny of the hybrids were grown to flowering. Among forty-four analyzed progeny with a *B. campestris*–like morphology, a few plants were found with the chromosome number of *B. campestris,* high pollen fertility, and cross-compatibility with pure *B. campestris* (Mikkelsen et al. 1996).

When wild *B. campestris* × cultivated *B. napus* hybrids were grown with their parents under field conditions, the hybrids were significantly more fit than wild *B. campestris,* producing, on average, fewer seeds per fruit but many more fruits per plant (Hauser et al. 1998b). Backcrosses and F_2s were also compared with wild *B. campestris* and cultivated *B. napus* under field conditions. These genotypes had, on average, reduced fitness relative to their wild grandparent (Hauser et al. 1998a). The authors of the study note, however, that some of the individuals were as fit as their parents, and conclude that the introgression of crop alleles from oilseed rape to wild *B. campestris* will be slowed, but not stopped, by the low average fitness of the second-generation hybrids. These conclusions are supported by the introgression observed by Hansen et al. (2001) above.

Spontaneous hybridization between *B. napus* and wild *B. juncea* in Denmark was investigated by planting twelve individuals of the latter in a field of the former (Jørgensen et al. 1996). Progeny from the wild plants were analyzed for hybridity with RAPDs, isozymes, chromosome counts, and morphological measurements. Three percent of the *B. juncea* progeny tested were identified as interspecific hybrids; Bing et al. (1996) reported the same hybridization rate for a *B. juncea* cultivar interplanted with *B. napus* oilseed rape in Canada. In a subsequent experiment wild *B. juncea* was planted at a range of frequencies in stands of cultivated *B. napus* in Denmark (Jørgensen et al. 1998). When *B. napus* was the maternal parent, the hybridization rate was constant, about 1 percent; when *B. juncea* was the maternal parent, the hybridization rate decreased from 2.3 percent to 0.3 percent as *B. juncea* became more frequent in the stand (increasing from 75 percent to over 90 percent).

Field-based experimental studies in France have also demonstrated that *B. napus* can act as a successful pollen parent in spontaneous intergeneric crosses, albeit at a very low rate. Herbicide-resistant *B. napus* was interplanted with the pantemperate weed hoary mustard *(Hirschfeldia incana = B. adpressa)* in a field at a low density (Lefol et al. 1996b). About 2 percent of seedlings from the wild plants were herbicide resistant. Isozyme, morphological, and cytogenetic analysis confirmed their hybridity. The *Hirschfeldia × Brassica* hybrids were all triploids, producing no fertile pollen grains and less than one seed per plant under greenhouse conditions. In a similar experiment, herbicide-resistant *B. napus* was interplanted with the serious pantemperate weed jointed charlock *(Raphanus raphanistrum)* under equal frequencies (1:1) and low frequency for the weed (1:600 charlock:rape) (Darmency et al. 1998). Progeny from the wild plants were tested for herbicide resistance and had their hybridity double-checked by isozyme and cytogenetic analysis. Only a few hybrids were detected in the low-frequency treatment in only one of the three years of study. The number of hybrid progeny per maternal plant averaged 0.03 over the whole experiment. No hybrids were detected in the equal-frequency treatment. An Australian field experiment involving the same species (Rieger et al. 2001) measured the frequency of hybridization to be extremely low. Not a single hybrid was found among twenty-five thousand seedlings from jointed charlock. Two hybrids were discovered among over fifty-two million seedlings from oilseed rape. *Raphanus × Brassica* hybrids have very low fertility under field conditions, averaging less than one seed per plant (Chèvre et al. 1998). Both paternal and maternal fertility in the field were observed to increase in first- and second-generation backcrosses to jointed charlock (Chèvre et al. 1998), although it is not clear from the data presented whether fitness in the second-generation backcrosses had recovered to the level typical of the wild species. A third French experiment involved a transgenic herbicide-resistant *B. napus* cultivar interplanted with the serious weed wild mustard *(Sinapis arvensis)* (Lefol et al. 1996a); despite progeny testing of millions of seeds from the wild species for herbicide resistance, no hybridization was detected.

Although *B. napus* can successfully cross spontaneously with certain species in other genera, it is reproductively isolated from certain *Brassica* species. For example, a field experiment in Saskatchewan failed to detect any spontaneous hybridization between *B. napus* and the weed *B. nigra* in hundreds of seeds that were tested (Bing et al. 1996).

#10. Dry, String, and Green Beans

The common bean, *Phaseolus vulgaris* L., comprises most of the world's production of dry, string, and green beans, which are primarily used by humans as a vegetable. This annual legume was apparently domesticated independently in both Mesoamerica and South America (Debouck and Smartt 1995). Most of the world's production comes from Asia and the New World. India is the most important producer, accounting for almost a quarter of the world's production. Four other *Phaseolus* species are cultivated to a relatively limited extent and will not be considered here.

Wild beans most closely related to the common bean are the wild forms of *P. vulgaris* that occur from Mexico through Central America and south along the Andes of South America into northern Argentina (Debouck and Smartt 1995). A number of other closely related wild taxa (including wild forms of the cultivated species *P. polyanthus*, *P. coccineus*, and *P. acutifolius*) also occur in Mesoamerica and South America (Debouck and Smartt 1995).

Plants appearing to be hybrids between cultivated bean and wild *P. vulgaris* often occur in South America at the margins of bean fields when wild beans are present (e.g., Beebe et al. 1997). These intermediates often persist for years (Freyre et al. 1996). Plants appearing to be hybrids between common bean and its wild relatives *P. aborigineus* and *P. mexicanus* are known from Mexico (Acosta et al. 1994, Delgado Salinas et al. 1988, Vanderborght 1983). However, morphological intermediates are not always present where cultivated beans and their wild relatives co-occur (Debouck et al. 1993, Freyre et al. 1996).

Genetic analysis of the wild intermediates in South America has been conducted using phaseolin seed proteins and progeny segregation studies of morphological traits. Intermediate plants have a combination of alleles specific to both the pure crop and the pure wild taxon (Beebe et al. 1997), supporting the hypothesis that the intermediates have a hybrid ancestry. In fact, some of the weedy populations are fixed for both crop-specific and wild-specific characters, suggesting that they are genetically stabilized lineages arising from past hybridization. I am not aware of similar genetic analysis of the putative hybrids in Mexico.

#11. Groundnut (Peanut)

The seeds of the annual legume *Arachis hypogaea* L. are eaten cooked in much of the world. They are also used as an oil crop, as a snack food, and as livestock feed and may be roasted and ground to make peanut butter. Although groundnut was domesticated in central South America, most of the world's production comes from Asia and Africa. China produces more than a third of the world's groundnuts.

Groundnut shares the same chromosome number with only one wild relative, the wild South American species *A. monticola* (both are tetraploids, 2n = 4× = 40 chromosomes). Experimental crosses between those species readily produce fertile hybrids. The crop does not cross freely with other wild relatives (which are largely diploid, 2n = 20 chromosomes) (Singh 1995). I did not find any report of spontaneous hybridization between groundnut and any wild relative. Although *A. hypogaea* may show limited natural cross-pollination (S. Hegde, personal communication), all *Arachis* species are considered to be highly self-pollinated (Singh 1995), a situation that would limit spontaneous interspecific hybridization, even in sympatry. And sympatry with wild relatives is extremely rare for groundnuts, because they are a minor crop in South America where *A. monticola* grows wild.

#12. Sunflower Seed

The annual sunflower, *Helianthus annuus* L., is an important oilseed plant. Its seeds are also eaten directly, both as human food and as animal fodder. Certain cultivars are grown primarily as ornamentals or for floricultural purposes. One of the few plant species domesticated in North America, sunflower is now a crop of the world's temperate regions, particularly Europe, South America, and Asia. Argentina leads the world in production, accounting for about 20 percent of the world's total.

The closest relative of cultivated sunflower is the wild form of the same species. Wild *H. annuus* is a weed in much of North America and has spread to other continents as well. Cultivated sunflower is cross-compatible with wild *H. annuus* and certain other wild *Helianthus* species, which are largely endemic to western North America (Rogers et al. 1982).

Both morphological and molecular evidence suggest that wild *H. annuus* naturally hybridizes with other wild sunflower species in areas of contact in western North America (e.g., Dorado et al. 1992, Heiser 1978, Rogers et al. 1982). Indeed, certain rare sunflower species naturally hybridize with wild (or possibly feral) *H. annuus,* which may increase their risk of extinction (Rogers et al. 1982). Molecular markers have confirmed the hybrid origin of certain *Helianthus* species involving *H. annuus* as one of the parent species (e.g., Rieseberg et al. 1990). However, the existing data on hybridization between wild taxa are of limited value in predicting gene-flow dynamics between cultivated *H. annuus* and wild congeners.

Spontaneous hybridization between cultivated sunflower and wild *H. annuus* has been the subject of considerable research. Allozyme progeny analysis of wild *H. annuus,* planted around cultivated sunflower fields in Mexico, detected spontaneous hybridization at substantial rates and over distances of up to one thousand meters from the crop (Arias and Rieseberg 1994). The fitness of hybrids between the wild and cultivated sunflower has been compared with that of wild plants under field conditions. "In general, hybrid plants had fewer branches, flower heads, and seeds than wild plants, but in two crosses fecundity of the hybrids was not significantly different from that of purely wild plants" (Snow et al. 1998). Sunflowers also provide one of the few cases of the relative field fitness of crop-wild hybrids under attack from biological enemies. Cummings et al. (1999) compared the seed produced by wild plants and wild-crop F_1s at three experimental field sites in eastern Kansas. "The average hybrid plant had 36.5% of its seeds . . . eaten by insect larvae while the average wild plant lost only 1.8% . . . to seed predators" (Cummings et al. 1999). The relative fitness of seeds produced by hybrids was examined under postdispersal conditions in the field (Alexander et al. 2001). In that experiment, "significantly more hybrid seeds were eaten (62% of the hybrid seed; 42% of wild seed)" (Alexander et al. 2001). But reduced fitness of the hybrids is not an absolute barrier to introgression. Introgression of crop alleles into wild sunflower populations is likely to be slowed rather than prevented by the lowered fitness of the hybrids.

Indeed, introgression from cultivated to wild *H. annuus* appears to occur with little impediment. Whitton et al. (1997) found crop-specific RAPD markers persisting in a California population of wild sunflowers for five generations following a single season of hybridization with a nearby crop plantation. Linder et al. (1998) analyzed three wild *H. annuus* populations from the northern Great Plains that had had long-term contact (20–40 years) with the crop. They found substantial introgression of crop-specific RAPD markers, with an "average overall frequency of cultivar markers greater than 35%," and "every individual . . . tested contained at least" one cultivar-specific allele. Given the substantial gene flow from the crop to its wild form, it is not surprising that introgression from the crop has been implicated in the evolution of increased weediness in wild *H. annuus* (Heiser 1978).

One study has addressed introgression of cultivated sunflower alleles into natural populations of another *Helianthus* species. Rieseberg et al. (1999) genetically analyzed four Great Plains populations of *H. petiolaris* that were distant from wild *H. annuus* populations but adjacent to fields of cultivated sunflowers. They found cultivar-specific AFLP alleles in a relatively small number of individuals and at relatively low frequencies within each population. "All hybrids appeared to represent later-generation backcrosses" (Rieseberg et al. 1999).

#13. Potato

The herbaceous perennial *Solanum tuberosum* L. is responsible for most of the world's potato production. While the tubers of this crop are usually cooked (with or without subsequent processing) for human consumption, they are also used for livestock feed or processed for flour, alcohol, and starch. Potatoes were domesticated in the high valleys of the South American Andes but are now grown throughout the temperate and upland tropical areas of the world. The vast majority of the world's production comes from Europe and Asia. China is the leading potato-producing country, accounting for about 15 percent of world production. *S. tuberosum* ssp. *tuberosum* L., an allohexaploid ($2n = 6\times = 72$ chromosomes), is grown worldwide; *S. tuberosum* ssp. *andigena* Juz. et Bu., an allotetraploid ($2n = 4\times = 48$ chromosomes), is cultivated mainly in the mountains of Mexico, Central America, and South America (Hanneman 1994). Several other potato-producing *Solanum* species are grown but are of minor importance globally. They will not be considered here.

Wild potatoes are an array of *Solanum* species ranging from the southwestern United States through Central and South America into southern Chile (Simmonds 1995). Both wild and cultivated potatoes comprise the tuber-bearing *Solanum* species, about two hundred species in a genus of thousands. Most tuber-bearing *Solanum* species are cross-compatible. Traditionally, natural hybridization has been thought to play a major role in the evolution of *S. tuberosum* and its close relatives, both wild and cultivated, in South America (e.g., Hawkes and Hjerting 1989), but that view is controversial (e.g., Spooner and Van den Berg 1992).

S. tuberosum is compatible with many of its wild relatives in Latin America, including *S. albicans, S. gourlayi* ssp. *gourlayi, S. moscopanum, S. oplocense, S. sucrense,* and seven species of *Solanum* series Demissa (D. M. Spooner, personal communication; Hanneman 1994). Furthermore, many diploid tuber-bearing *Solanum* species that occasionally produce unreduced (2n) gametes can be crossed with the tetraploid *S. tuberosum* ssp. *andigena* (den Nijs and Peloquin 1977; Mok and Peloquin 1975). However, *S. tuberosum* is strongly incompatible with wild tuber-bearing *Solanum* species of the United States and Canada (Love 1994).

The closest wild relatives of *S. tuberosum* are members of the same species that have become naturalized in South America (Hawkes 1990). Although strong reproductive barriers between naturalized and cultivated *S. tuberosum* are unlikely, spontaneous hybridization rates between the two forms have apparently not been measured.

Potatoes are usually harvested prior to flowering. They are propagated vegetatively by tuber cuttings. Many potato cultivars are fertile; others are fully or partially sexually sterile. While that sterility may limit opportunities for hybridization, potato's perenniality and ability to reproduce vegetatively may foster the persistence and spread of hybrids even if their sexual fertility is reduced (cf. Ellstrand et al. 1996).

A number of spontaneous hybrids between *S. tuberosum* and wild relatives have been reported. *S. sucrense,* which grows primarily as a weed of cultivation in Bolivia, may be a hybrid of *S. tuberosum* ssp. *andigena* and the wild *S. oplocense* (Astley and Hawkes 1979). Its hybridity has been supported by the fact that synthetic hybrids of the two putative parents are extremely similar to the wild plant. Also, seedlings collected from certain *S. sucrense* populations in Bolivia segregate for morphological characters associated with the parental species. However, cytogenetic analysis of *S. sucrense* suggests that it is an autotetraploid without a history of hybridization (Ochoa 1990). The Bolivian weed potato, *S. × subandigena,* is apparently a stabilized hybrid of *S. tuberosum* ssp. *andigena* and *S. sucrense,* but the assignment of hybridity has been based solely on morphology (Hawkes 1990).

Spontaneous hybrids of *S. tuberosum* and *S. demissum* are named *S. × edinense* (Hawkes 1990). Their hybridity has been supported by cytogenetic analysis, which revealed the pentaploid chromosome number expected for hybrids between hexaploid and tetraploid taxa. *S.* × *edinense* ssp. *edinense* appeared at the Edinburgh Botanic Gardens with *S. tuberosum* ssp. *tuberosum* as the crop parent. *S.* × *edinense* ssp. *salamanii,* a weed of potato fields in central Mexico, is a natural hybrid of *S. tuberosum* ssp. *andigena* and *S. demissum* (Hawkes 1990).

However, thorough population-level genetic analysis has not yet been conducted to determine the extent to which crop alleles have moved or persist in natural populations. For example, *S.* × *edinense* ssp. *salamanii* could be either a stabilized hybrid-derived taxon or simply a name for occasional seed-sterile hybrids of *S. tuberosum* ssp. *andigena* and *S. demissum* that persist through vegetative reproduction.

Isolation barriers are very strong between *S. tuberosum* and the thousands of non-tuber-bearing *Solanum* species. Field experiments have been conducted to test whether *S. tuberosum* will spontaneously mate with two non-tuber-bearing species that are noxious weeds, *S. nigrum* and *S. dulcamara.* At an English study site, McPartlan and Dale (1994) planted both weed species around a plot of potatoes transgenic for kanamycin resistance. Thousands of seedlings from the weeds were screened; not one showed resistance to the antibiotic, demonstrating that the barriers to hybridization between potato and these weeds are indeed very strong.

#14. Sugarcane

Several large perennial grass species in the genus *Saccharum* comprise sugarcane. Sugar is processed from the juice of their stems. Noble cane, *S. officinarum* L., accounts for most of the world's production of cane sugar. This species, now found only in cultivation, was apparently domesticated in New Guinea (Roach 1995). A tropical crop, it is particularly important in South America and Asia; Brazil ranks first in production, producing over 25 percent of the world's crop.

The closest wild relatives of *S. officinarum* are an array of *Saccharum* species that occur mostly in South and East Asia, Indonesia, and New Guinea (Roach 1995). One relative, *S. spontaneum,* is cross-compatible with the crop and frequently used in sugarcane breeding programs (Roach 1995). That species is also a serious weed of tropical Asia and of some sporadic concern in Africa, Europe, and the tropical New World (Holm et al. 1997). Noble cane is also cross-compatible with other Old World genera, in particular certain species of *Erianthus, Miscanthus, Narenga,* and *Sclerostachya* (Roach 1995).

Sugarcane does not need to flower to produce its crop; it is propagated vegetatively by stem cuttings. Its perenniality and ability to reproduce vegetatively may foster the persistence and spread of hybrids, even if their sexual fertility is reduced (cf. Ellstrand et al. 1996).

Spontaneous hybridization between cultivated and wild *Saccharum* in Australasia and some islands of the Indian Ocean has been implicated in the origin of many cultivars (Roach 1995). While evidence for the hybrid origin of these cultivars comes most frequently from their morphology, some supporting evidence also comes from genetically based traits. For example, certain wild plants from New Guinea that are morphologically classified as *S. spontaneum* "have leaf flavonoids and triterpenoids common to . . . *S. officinarum*" (Daniels and Roach 1987). Likewise, chromosome data indicate that certain wild canes from Java and Mauritius are hybrids between the cultivated *S. officinarum* and the wild *S. spontaneum* (Stevenson 1965).

Spontaneous intergeneric hybridization has been implicated in the origin of particular accessions of wild sugarcane relatives (Daniels and Roach 1987). Chromosomal evidence supports the suggestion, based on morphology, that *Erianthus maximus* is derived from a natural cross of the cultivated *S. officinarum* with the wild genus *Miscanthus* (Daniels and Roach 1987). Likewise, a *Miscanthus* clone was found to have the same chloroplast restriction fragment sites as cultivated *Saccharum* species, suggesting a history of introgression between the two genera in the lineage of that genotype (Sobral et al. 1994). Population-level analysis has not yet been conducted to determine the extent to which crop alleles have moved into natural populations.

#15. Cassava

Cassava (also known as manioc or yuca) is the perennial shrub *Manihot esculenta* Crantz. Tubers of this crop are usually cooked (with or without subsequent processing) for human consumption. They are also processed for tapioca, alcohol, and industrial starch or used for livestock feed. The leaves are eaten as a potherb. Cassava was domesticated in South America, but the specific location remains controversial. It is now grown throughout the world's wet lowland tropics. More than two-thirds of the world's production comes from Africa and South America. The leading cassava-producing country is Nigeria, accounting for about 20 percent of world production.

Cassava need not flower to produce its crop and is propagated vegetatively by stem cuttings. Cassava's perenniality and the ability of at least some *Manihot* hybrids to produce seeds asexually (Nassar et al. 1998) may foster the persistence and spread of hybrids even if their sexual fertility is reduced (cf. Ellstrand et al. 1996).

Wild *Manihot* species are native from the southwestern United States south to Argentina. Some have been introduced elsewhere (e.g., *M. glaziovii;* see below). All species can be intercrossed (Jennings 1995, Nassar 1980). The great variability within the crop has been attributed to hybridization with wild relatives (Rogers and Appan 1973; N. Nassar, personal communication), but little is known of the movement of crop alleles into natural populations. A large number of "wild species have a propensity for colonizing disturbed areas and there would have been ample opportunity for gene exchange" with the crop "in areas adjacent to cultivation" (Jennings 1995). In fact, the range expansion of *M. reptans* in the last century has been attributed to continuing introgression from cassava; plants of a weedy population of *M. reptans* were found to bear certain cassava-like morphological traits (Nassar 1984).

A putative hybrid swarm between cassava and a wild relative in the Ivory Coast was genetically examined with allozymes (Lefèvre and Charrier 1993). Interestingly, the wild species was *M. glaziovii,* which had become naturalized in Africa after being introduced about sixty years earlier for unsuccessful rubber production experiments. Most of the putative hybrids were heterozygous for alleles specific to both cassava and *M. glaziovii,* confirming their hybridity. But two accessions were homozygous for cassava-specific alleles at a few loci, suggesting that they were probably spontaneous backcrosses to the crop. Studies of this kind would be helpful in testing the hypothesis that cassava hybridizes with wild *Manihot* species where they are sympatric in the New World.

#16. Oats

Although oats, the annual grass *Avena sativa* L., is a major temperate cereal crop, its importance is gradually declining. The grain is typically processed and cooked for human consumption; both the grain and straw are used for livestock feed. It is not certain where oats were domesticated, but the site is likely to be in Europe (Thomas 1995). More than half of the world's production comes from temperate North America, Europe, and Russia (the latter being the world's primary producer, accounting for about 16 percent of the production of the crop).

Experimental crosses between *A. sativa* and its closest wild relatives, *A. fatua* and *A. sterilis*, are easily carried out; cultivated oat is not compatible with other wild *Avena* species (Thomas 1995). Both *A. fatua* and *A. sterilis* are native to Europe, North Africa, the Middle East, and Central Asia (Baum 1977). Plants of disturbed areas, their ranges have been extended through inadvertent human assistance to become two of the world's worst temperate weeds (Baum 1977, Holm et al. 1977).

Plants with intermediate morphologies between oats and *A. fatua* "usually appear sporadically in or around oat-fields or where oats have been previously grown . . . across Europe and throughout much of Canada" as well as the United States and Japan but are rarely persistent (Stace 1975a, Baum 1977). Because synthetic hybrids have morphologies that are quite similar to the domesticated parent, "natural intermediates are more likely to be backcrosses or segregants than F_1 hybrids, and some are possibly mere variants of *A. fatua*" (Stace 1975a). Despite morphological evidence for hybridization, I am not aware of any genetic analysis of the spontaneous intermediates.

However, the spontaneous intermediates are not the same as "fatuoids," *fatua*-like plants arising in populations of cultivated oats. Fatuoids were once thought to arise as products of the segregation of spontaneous *A. sativa* × *A. fatua* hybrids. However, the current view is that fatuoids arise by chromosomal mutation (Baum 1977).

Burdon et al. (1992) tested hundreds of seeds set in four naturally mixed Australian populations of *A. sativa* and *A. fatua* for species-specific isozyme alleles. That study revealed the spontaneous hybridization rate to be about 0.7 percent. In England, Derick (1933) created experimental plots of *A. fatua* surrounded by *A. sativa*. Thousands of crop progeny assayed for the dominant dark lemma color specific to the wild species revealed that 0.1 to 0.31 percent of those plants were interspecific hybrids. Segregation analysis of the putative hybrids confirmed their hybridity.

Plants with intermediate morphologies between oats and *A. sterilis* have been reported from Europe, the Middle East, and Central Asia "where the two species co-exist" (Baum 1977). Despite morphological evidence for hybridization, I am not aware of any genetic analysis of the spontaneous intermediates.

Nor am I aware of experiments to assess natural hybridization rates between oats and *A. sterilis*.

The fitness of either *A. sativa* × *A. fatua* hybrids or *A. sativa* × *A. sterilis* hybrids is expected to be low in field situations because the hybrid "will resemble the cultivated parent more closely," including those traits that are handicaps in natural populations, such as nondormancy (Barr and Tasker 1992). The fitness of the F_2s is expected to be low for the same reason; only about 1 percent of the F_2s will recover all of the most important weedy traits (Barr and Tasker 1992). Still, under constant gene flow, the observed hybridization rates and hybrid fitnesses would not prohibit neutral or beneficial crop alleles from entering and persisting in natural populations (Barr and Tasker 1992; see chap. 4).

#17. Coconut

The coconut palm, *Cocos nucifera* L., is a tree of the lowland humid tropics. Its primary commercial product is the oil that is extracted from the (fresh or dried) endosperm of its ripe fruit. However, the tree has an amazing array of other uses; of the crops covered in this review, only maize may be equally versatile.

The mature endosperm can be eaten directly or grated for use in baked goods. Fibers from the mesocarp (coir) are used for stuffing, netting, and brushes. Coir dust (cocopeat) is a sustainable replacement for horticultural peat moss. The shell can be used to produce high-quality activated charcoal. Coconut water, the liquid in the fruit, may have been the original motivator for the plant's domestication, as a source of pure water. Coconut milk and coconut cream, emulsions of coconut oil in water, are essential ingredients in curries and certain other prepared foods. The coconut meal residues from commercial processing go into animal feeds. Uses for the plant extend beyond the fruit. For example, leaves can be woven and used for house walls and thatch, the sap from the inflorescence can be fermented into alcohol or boiled and crystallized into syrup or sugar, the trunk can be used to produce "porcupine wood" timber, and the "heart" can be eaten as a vegetable (H. Harries, personal communication).

The coconut palm is indigenous to shores along both the Indian and Pacific Oceans (Harries 1995). The center of domestication for this species is still controversial; it is likely to be in or near Southeast Asia. Both domesticated and wild coconut palms have been introduced worldwide into suitable habitats (Harries 1995). Asian countries account for well over three-quarters of the world's production. The Philippines accounts for about one-quarter of the world's total production.

The genus *Cocos* is monotypic. Thus, the only wild compatible relatives of domesticated coconut are wild and feral members of the same species. Domesticated coconut is fully cross-compatible with other coconuts, wild or feral (Harries 1995). Spontaneous hybridization appears to occur wherever domesticated trees co-occur with those that are free-living. Because of this introgression, no truly wild coconuts may be left (Harries 1995).

I am not aware of any study testing hypotheses of spontaneous hybridization between wild and domesticated coconut palms using genetic markers. However, almost any coconut tree may be suitable for cultivation (H. Harries, personal communication). If trees are essentially interchangeable, then genetic markers might not be available to distinguish "domesticated" trees from those with "pure" wild ancestry. Thus, genetically based identification of hybrids and introgressants might be impossible for coconut. However, two RAPD markers have recently been identified that may prove useful. One marker occurs at a very low frequency in wild material and is extremely common in trees with a substantial number of traits associated with a long history of domestication; the other marker has the opposite pattern (Ashburner and Harries 1999).

#18. Coffee

Coffee, the world's most important beverage stimulant, is brewed from the roasted and ground seeds of trees in the genus *Coffea*. Well over 90 percent of the world's commercial production comes from two species, arabica coffee, *C. arabica* L., and robusta coffee, *C. canephora* Pierre ex Froehner (Wrigley 1995a). The most important commercial species, arabica coffee, is grown throughout the world's tropical highlands. It is indigenous to southwestern Ethiopia and adjacent Sudan, where a few wild populations still exist (Wrigley 1988). *C. arabica* is an allotetraploid derivative of a hybrid of *C. congensis* and *C. eugenioides* (Raina et al. 1998). Robusta coffee is cultivated primarily in the world's tropical lowlands but also grows wild from West Africa east to Uganda and Tanzania and south to Angola. But it is not clear whether that is the native range of *C. canephora* or whether its range was extended under prehistoric cultivation (Wrigley 1988). All other wild coffee species are indigenous to Africa, Madagascar, and the Mascarenes. Coffee is produced throughout the world's tropics, but production is particularly high in South America. Brazil is the top producer, accounting for about one-quarter of the world's production.

Experimental crosses between the tetraploid ($2n = 4\times = 44$ chromosomes) *C. arabica* and the wild species in the genus, which are all diploid, rarely produce progeny, and those progeny are typically triploid and sterile (Van der Vossen 1985). Nonetheless, the morphological characters of several cultivars (but a small fraction of the total) suggest that they may be spontaneous hybrids between *C. arabica* and relatives such as *C. liberica*, *C. canephora*, *C. dewevrei*, and *C. stenophylla* (Wrigley 1988). *C. dewevrei* and *C. stenophylla* exist only as wild species, but in the case of *C. liberica* and *C. canephora* the pollen parent might have been cultivated, feral, or wild. In a few cases cytogenetic analysis has confirmed the hybrid parentage of these morphologically intermediate cultivars (Wrigley 1988).

In a few cases other genetic markers have been used to identify hybrids between wild *Coffea* species and *C. arabica*. RAPD markers specific to *C. canephora* were identified in a *C. arabica* cultivar, Rume Sudan (Orozco-Castillo et al. 1994). This cultivar "was identified in seed collected from wild coffee growing . . . in one of the few regions where *C. arabica* [and] *C. canephora* . . . coexist" (Orozco-Castillo et al. 1994). Isozyme analysis of Piãta coffee revealed a combination of alleles specific to *C. arabica* and the wild species *C. dewevrei,* supporting morphological and cytogenetic evidence of hybridity (Medina Filho et al. 1995, detailed in chap. 5 above). Clearly, *C. arabica* has spontaneously hybridized with its wild relatives, but I have not found any study that has addressed whether alleles from cultivated *C. arabica* have introgressed into natural populations.

Experimental crosses between *C. canephora* and wild *Coffea* species vary widely in success (Charrier and Berthaud 1985). For example, *C. canephora* crosses readily with *C. congensis,* producing fertile progeny; in contrast, progeny from crosses with *C. resinosa* have very low fertility. The morphological characters of a few cultivars suggest that they may be spontaneous hybrids between *C. canephora* and its wild relatives *C. congensis* and *C. excelsa* (Wrigley 1988). These are the only data suggesting that *C. canephora* has spontaneously hybridized with its wild relatives. I am not aware of any genetic analysis of the spontaneous intermediates. Nor have I found any study that has addressed whether alleles from cultivated *C. canephora* have introgressed into natural populations.

#19. Chickpea

The seeds of the annual legume chickpea, *Cicer arietinum* L., are used in a variety of ways for human consumption: eaten fresh as green vegetables or sprouted, cooked, and processed for flour. The seeds and other parts of the plant are also used for livestock feed. Although chickpea was domesticated in southeastern Turkey, it has been widely distributed (Ladizinsky 1995). Most of its current production is in Asia, especially India, which produces more than two-thirds of the world's crop.

Experimental crosses between chickpea and its wild relative *C. reticulatum*, which is known from only a few populations in southeastern Turkey, readily produce highly fertile hybrids. Crosses between chickpea and another wild relative, *C. echinospermum* of the Middle East, can be conducted with some difficulty and produce partially or fully sterile hybrids. The crop does not cross freely with other *Cicer* species (Ladizinsky and Adler 1976, Ahmad et al. 1988). I did not find any report of spontaneous hybridization between chickpea and any wild relative. However, all species of *Cicer* are presumed to be highly self-pollinated, which may limit interspecific gene flow. Also, the extremely restricted geographic distribution of the most compatible wild relative limits the opportunities for hybridization with the crop.

#20. Cowpea (Black-Eyed Pea)

The annual legume *Vigna unguiculata* ssp. *unguiculata* (L.) Walp. produces cowpeas. Plant parts that are eaten include dry seeds, green pods, green seeds, and tender green leaves. As recently as 1995 it was characterized as a "crop in the subsistence agriculture of the semiarid and subhumid tropics of Africa" (Ng 1995). At that time, it ranked thirtieth in total area planted. But the crop is gaining steadily in importance, as evidenced by the fact that it now ranks twentieth. Most of the world's production is still in Africa, with over two-thirds coming from Nigeria alone. However, the crop is of importance elsewhere, especially Brazil, India, and (as the "yard-long bean" type) certain areas of southern China and Southeast Asia.

The crop and many of the wild subspecies of *V. unguiculata,* which are endemic to sub-Saharan Africa, are interfertile, bearing fertile hybrids. The species is not compatible with other members of the genus (Rawal 1975, Ng 1995). Companion weeds with a range of morphologies between cowpea and related wild conspecifics are found along roadsides as well as in disturbed areas in Niger and Nigeria (Rawal 1975). Despite morphological evidence for hybridization, I am not aware of any genetic analysis of the spontaneous intermediates.

#21. Rye

Rye, the annual grass *Secale cereale* L., is the most cold-tolerant temperate cereal crop, typically grown in environments too harsh for wheat. The grain is processed into flour or fermented and distilled in the production of alcoholic beverages. The plants are also used for livestock forage. Rye was probably domesticated from wild plants that were companion weeds of barley and wheat in the Near East (Evans 1995). About two-thirds of the world's production comes from Europe. Russia is currently the world's leader, producing about 20 percent of the crop.

Cultivated rye and the wild subspecies of *S. cereale* are completely interfertile. These weeds occur from the Near East east to Iran, Afghanistan, and Transcaspia (Evans 1995). Plants with morphologies intermediate to rye and its related wild conspecifics have been noted where they co-occur in central and eastern Turkey (Zohary 1971). However, I am not aware of any genetic analysis of these spontaneous intermediates.

Cultivated rye and the perennial species *S. montanum* are also interfertile (Evans 1995). *S. montanum* is a highly variable taxon; its native range is distributed widely from the Mediterranean basin east through Turkey to northern Iraq and Iran. It has been introduced as a source of forage into temperate regions, including Australia, New Zealand, and North America. Apparent hybrid swarms of plants with morphologies intermediate to rye and *S. montanum* occur sporadically where the two species co-occur on the Anatolian plateau (Zohary and Hopf 1993). The existing data are apparently based on morphology alone. Genetic analysis is necessary for testing whether these are indeed hybrid swarms.

An aggressive weedy rye appeared in northeastern California in 1964. It has some characteristics of cultivated rye, others specific to *S. montanum,* and is intermediate for other characters, suggesting that it is a hybrid derivative of *S. cereale* and *S. montanum* (Suneson et al. 1969). Genetic analysis of this weed supports its hybrid origins. Seed families from weeds segregate for traits specific to the parental species (Jain 1977). Isozyme analysis of the weedy California populations lends further support with alleles that "were shared with either" or both a common rye cultivar "and accessions of *S. montanum*" (Sun and Corke 1992). Both putative parental species are known to have been growing in this region, permitting opportunities for hybridization. Rye was cultivated as an important crop. "Michel's grass," a derivative of *S. montanum,* was introduced into the region in 1938. Natural hybridization probably occurred soon after that. Traits specific to *S. cereale* have increased over time in the wild populations, probably as a result of a combination of continuing gene flow from the crop and natural selection (Jain 1977). The hybrid derivative's fitness in the field (in terms of plant biomass and seed production) is superior to that of its wild progenitor (Suneson et al. 1969). Within a few decades of the discovery of weedy rye in California, "this introgression . . . [had] proceeded to such an extent that farmers [in this region] . . . abandoned efforts to grow cultivated rye [and wheat] for human consumption" (National Academy of Sciences 1989; Suneson et al. 1969). "Further spreading of the weedy populations seems to be in progress" (Sun and Corke 1992), extending even into southern California (Hickman 1993).

#22. Oil Palm Fruit

Oil palm fruits come from the tree *Elaeis guineensis* Jacq. The fruits are processed for two products: palm oil from the mesocarp and kernel oil from the seed. After oil extraction, the remaining kernel is used for animal feed. In less than half a century, palm oil and kernel oil have progressed from obscure products to ranking among the most important vegetable oils in the world. Although *E. guineensis* is native to West Africa, it has been introduced into suitable regions of the New World and Asia. The great bulk of the world's production is now in Asia, particularly in Malaysia, which accounts for about 43 percent of the total.

The closest wild relatives of *E. guineensis* are feral or wild populations of the same species in the New World and "throughout the rain forest belt of West Africa, from Cape Verde to Angola" (Hardon 1995). Given that oil palms grown in Africa are "little, if at all, removed from the wild type" (Hardon 1995), it is likely that spontaneous hybridization occurs freely between the crop and wild populations of the same species in these regions. However, I am not aware of any study that documents such hybridization.

The genus *Elaeis* includes two other species (Hardon 1995), both of which are wild. *E. oleifera* is endemic to northern South America; its range extends north to Costa Rica. This species hybridizes readily with *E. guineensis* to produce offspring that are fertile but have reduced pollen viability (Hardon and Tan 1969). "In Colombia and Costa Rica natural hybrids have been found when *E. oleifera* palms grew in close proximity to commercial plants of *E. guineensis*" (Meunier and Hardon 1976). Despite morphological evidence for spontaneous hybridization, I am unaware of any genetic analysis of the spontaneous intermediates. The third species in the genus, *E. odora,* is little known, distributed in scattered populations in Amazonia. I am not aware of any data regarding hybridization between the crop and this species.

#23. Sweet Potato

Sweet potatoes are roots of the herbaceous perennial *Ipomoea batatas* (L.) Lam. A perennial in the tropics, the crop grows as an annual in temperate climates. While sweet potatoes can be cooked for human consumption, they are frequently used for livestock feed or processed for starch, wine, and alcohol. Sweet potatoes were probably domesticated in Mesoamerica or northern South America (Bohac et al. 1995) but are now grown throughout the warm temperate and tropical areas of the world. The vast majority of the world's production comes from Asia. China is the leader, accounting for 85 percent of the total.

Wild sweet potatoes are an array of about ten *Ipomoea* species (Bohac et al. 1995). With *I. batatas,* they comprise the section Batatas. The wild species are largely confined to the New World, ranging from North America through Central America and the Caribbean into South America. However, one species occurs on shores along the Pacific and Indian Oceans.

Most wild species in the section are diploids (2n = 30 chromosomes); the cultivated species occurs both as tetraploid (2n = 4× = 60 chromosomes) and hexaploid (2n = 6× = 90 chromosomes) races. The differences in chromosome number create a reasonably strong barrier to interspecific hybridization. *I. trifida* of Central and South America is the only wild species that is able to produce fertile progeny when crossed with the crop (Diaz et al. 1996). One researcher has suggested that certain weedy tetraploid *Ipomoeas* are the result of spontaneous hybridization between sweet potato and *I. trifida* (Austin 1977).

However, the prevailing view is that the two species do not spontaneously hybridize in the field (D. LaBonte and J. Eckenwalder, personal communication). The question of whether sweet potato and *I. trifida* hybridize under field conditions clearly needs further study.

#24. Olive

Olives are fruits of the tree *Olea europaea* ssp. *europaea* L. The fruits are used in two ways: processed for olive oil and preserved for human consumption. The cultivars are typically propagated vegetatively by any one of a number of techniques. The olive was domesticated in the Mediterranean basin (Zohary 1995). That region remains the major olive-producing area of the world, especially Spain, which accounts for more than one-quarter of the world's production.

The closest wild relatives of *O. europaea* ssp. *europaea* are apparently feral populations of the same subspecies (Zohary and Hopf 1993). It is likely that spontaneous hybridization occurs freely between the crop and feral populations of the same subspecies in these regions (cf. Ouazzani et al.1993). However, I am not aware of any study that documents such hybridization.

Olive's wild progenitor, *O. europaea* ssp. *oleaster,* is endemic to the Mediterranean basin (Zohary 1995), from southern Portugal and western Morocco east to Israel and Lebanon. "It is fully interfertile with the crop and interconnected with it by sporadic hybridization" (Zohary 1995). Despite morphological evidence for spontaneous hybridization, I am not aware of any genetic analysis of the spontaneous intermediates. Hybridization has been implicated in the evolution of morphologically intermediate weedy types of *O. europaea* (Zohary and Hopf 1993). But without genetic analysis, it is impossible to know whether such weedy types are hybrid derivatives, the products of convergent evolution, or simply the sexual recombinants of cultivated clones.

Olive cultivars are also probably interfertile with several non-Mediterranean wild olives (Zohary 1995). I am not aware of any reports of spontaneous hybridization involving the crop and any of these wild species.

#25. Grape

The woody vine *Vitis vinifera* L. produces the great majority of the world's grapes. Its fruits are eaten fresh or dried. They are also processed for wine,

Table 7.3 Spontaneous Hybridization between the World's Most Important
Food Crops and Their Wild Relatives

Crop	Name	Evidence[a] for Hybrid-ization	Geographic Distribution of Hybrid-ization	Some Hybridizing Relatives
Wheat	*Triticum aestivum*[b]	+	Multicontinental	Certain *Triticum*, *Aegilops* taxa
	T. turgidum durum[b]	+	Multicontinental	Certain *Triticum*, *Aegilops* taxa
Rice	*Oryza sativa*	+	Multicontinental	*O. sativa* f. *spontanea*, *O. nivara*, *O. rufipogon*
	O. glaberrima	+	Africa	*O. barthii*, *O. longi-stamina*
Maize (sweet and field corn)	*Zea mays mays*	+	Mexico	Wild *Z. mays* subspecies, certain *Zea* species
Soybean	*Glycine max*	+	N.E. Asia	*G. soja*
Barley	*Hordeum vulgare*	m	W. Asia	*H. spontaneum*
Sorghum	*Sorghum bicolor bicolor*	+	Multicontinental	Wild *S. bicolor* sub-species, certain *Sorghum* species
Millet	*Eleusine coracana coracana*[b]	m	Africa	*E. coracana africana*
	Pennisetum glaucum[b]	+	Africa	*P. violaceum*, *P. sieberanum*
Cottonseed	*Gossypium barbadense*	+	Mesoamerica, Caribbean	Feral, wild *G. hirsutum*
	G. hirsutum	+	Multicontinental	Wild *G. barbadense*, certain *Gossypium* species
Rapeseed (canola)	*Brassica napus*	+	Europe	Wild *B. campestris*, *B. juncea*, *Hirschfeldia incana*, *Raphanus raphanistrum*
	B. campestris	+	Experimental results only	Wild *B. campestris*
Beans, dry, green and string	*Phaseolus vulgaris*[b]	+	Multicontinental	Wild *P. vulgaris*, certain *Phaseolus* species
Groundnut (peanut)	*Arachis hypogaea*	None	n/a	n/a
Sunflower seed	*Helianthus annuus*	+	North America	Wild *H. annuus*, *H. petiolaris*
Potato	*Solanum tuberosum*[b]	+	Multicontinental	Certain *Solanum* species

Table 7.3 Continued

Crop	Name	Evidence for Hybrid-ization[a]	Geographic Distribution of Hybrid-ization	Some Hybridizing Relatives
Sugarcane	*Saccharum officinarum*[b]	+	Australasia, Indian Ocean	Certain *Saccharum*, *Erianthus, Miscanthus* species
Cassava	*Manihot esculenta*	+	Multicontinental	*M. glazovii, M. reptans*
Oats	*Avena sativa*	+	Multicontinental	*A. fatua, A. sterilis*
Coconut	*Cocos nucifera*	m	Multicontinental	Feral, wild *C. nucifera*
Coffee	*Coffea arabica*[b]	+	Africa	Certain *Coffea* species
	C. canephora[b]	m	Africa	Certain *Coffea* species
Chickpea	*Cicer arietinum*	None	n/a	n/a
Cowpea	*Vigna ungui-culata ungui-culata*	m	Africa	Wild *V. unguiculata* subspecies
Rye	*Secale cereale*	+	Multicontinental	*S. montanum*
Oil palm fruit	*Elaeis guineensis*	m	Costa Rica, Colombia	*E. oleifera*
Sweet potato	*Ipomoea batatas*	?	Mesoamerica?	*I. trifida?*
Olive	*Olea europaea europaea*	m	Mediterranean basin	*O. europaea oleaster*
Grape	*Vitis vinifera*[b]	+	Multicontinental	Many *Vitis* species

[a]m, morphological intermediacy only; +, more substantial evidence for hybridization; ?, controversial.

[b]Other taxa account for a small portion of the world production of this crop.

juice, jam, and grapeseed oil. *V. vinifera* was probably domesticated in Central Asia (Olmo 1995) but is now grown worldwide in a variety of habitats, both temperate and tropical. More than half of the world's grape production is from Europe. Italy is the world's leading producer, accounting for about 16 percent of the total. *Muscadinia rotundifolia,* domesticated in North America in the 1600s (Olmo 1995), also produces grapes, but only a tiny fraction of the world's total, and will not be considered here.

A number of wild *Vitis* taxa are potential spontaneous mates with cultivated grape. The progenitor of cultivated grape, wild *V. vinifera* (= *V. sylvestris*), still occurs in Central Asia and in isolated populations in southern Europe and North Africa (Olmo 1995). "The boundary between the cultivated grape clones and the wild forms is blurred by the presence of escapees and derivatives of hybridization. Spontaneous crossing between wild plants and cultivars has been found repeatedly where *sylvestris* vines grow in close proximity to vineyards; F_1 hybrids are fully fertile; so in *V. vinifera* we are faced in the Mediterranean basin with a variable complex of wild forms (growing in primary habitats, escapees and seed-propagated weedy types (which occur mainly in disturbed settings), and cultivated clones" (Zohary and Hopf 1993). Despite the morphological evidence for intraspecific hybridization, I am not aware of any genetic analysis of the spontaneous intermediates.

Dozens of wild *Vitis* species occur in both the New and Old Worlds, including tropical, subtropical, and temperate regions. "All known *Vitis* species can be easily crossed experimentally, and the F_1s are vigorous and fertile" (Olmo 1995). Spontaneous intermediates have occurred in numerous instances where cultivated grape is grown in close proximity with wild congeners, both within its range and at the fringes of its range (Olmo 1995). As *V. vinifera* was introduced into new environments beyond its natural range, these putative spontaneous hybrids with native *Vitis* often provided the start for new cultivated races better adapted to local conditions (Olmo 1995). Most of these putative hybrids have been recognized by morphology alone. However, spontaneous hybridization involving *V. vinifera* × *V. californica* and *V. vinifera* × *V. girdiana* in California has been confirmed by both progeny segregation studies and isozyme analysis (Olmo and Koyama 1980). Population-level analysis has not yet been conducted to determine the extent to which crop alleles have moved into natural populations.

Vitis does not hybridize naturally with *Muscadinia*. Hand crosses between the two genera result in sterile hybrids (Olmo 1995).

Conclusions

The foregoing review suggests that spontaneous hybridization between a given crop and at least one wild relative somewhere in the world is apparently the rule rather than the exception (see the summary in table 7.3). For the twenty-five crops reviewed, all but three have some evidence for natural hybridization with one or more wild relatives. The three exceptions are groundnut *(Arachis hypogaea)*, chickpea *(Cicer arietinum)*, and sweet potato *(Ipomoea batatas)*, and even the case for sweet potato is controversial. The remaining twenty-two crops represent twenty-eight species. Of these, the majority, twenty-one species, have some sort of substantial genetically based evidence supporting natural hybridization with wild relatives. For only seven species is morphology the sole basis for presuming hybridization.

The twenty-eight species for which there is some evidence of hybridization are biologically diverse to an amazing extent, representing twenty-three genera and eleven plant families (see table 7.1). Few generalizations emerge from this list. The species include annuals and perennials, herbaceous plants and woody ones, those grown worldwide and those largely restricted to a single continent. Temperate and tropical crops are represented. Agronomic and horticultural crops are on the list, including fruits, cereals, vegetables, pulses, and oil crops. Some of the species rely on insects as their pollen vectors; others, wind. Their breeding systems range from highly outcrossing to largely selfing, and some are capable of reproduction without sex (table 7.2). Even after removing those species for which morphology is the sole evidence for hybridization, the list remains very heterogeneous.

Even the geographic extent of hybridization varies from crop to crop but is never highly localized. At the minimum, hybridization, when reported, appears to have occurred at multiple locations, usually in multiple countries. In more than half the species surveyed (16), hybridization has been reported from a single continent or a relatively restricted region within a continent (table 7.3). For the other species, hybridization has been reported from at least two continents.

The taxonomic extent of hybridization also defies simple generalizations (table 7.3). Five of the crop species surveyed appear to hybridize only with wild members of the same species. But the majority engage in interspecific hybridization. Interestingly, four of the species on the list not only naturally hybridize with other members of the same genus but engage in intergeneric hybridization as well. But this pattern may have more to say about the taxonomic assignment of the species involved than about their biological relationships, since assignment to genus is more arbitrary than assignment to a species.

It is equally hard to make generalizations for the three species for which spontaneous hybridization does not seem to occur. Two, groundnut and chickpea, are highly selfing annuals in the pea family (Fabaceae). They also have the option of cross-pollination by insects. But sweet potato is an outcrossing perennial in the morning glory family (Convolvulaceae). The only apparent common thread among the three is that they are all insect-pollinated.

In summary, almost 90 percent of the world's twenty-five most important crops naturally hybridize with wild relatives. This fraction is essentially the same as that found in my earlier collaborative study of the world's thirteen most important crops (Ellstrand et al. 1999), and indeed, the evidence for many of those previously surveyed is now considerably stronger. Overall, the data suggest that spontaneous hybridization with wild relatives is a general feature not only of plants grown for human consumption but also of domesticated plants in general. That is the topic of the next chapter.

CHAPTER 8

Is Natural Hybridization with Wild Relatives the Rule for Domesticated Plants?

Gene escape is a fact of nature. Period.
—ALAN MCHUGHEN, PLANT GENETICIST, 2000

Breeders have found that, with rare exceptions, the crops do not successfully cross-breed with other plants in the environment, especially plants in crop-growing regions.
—MARTINA MCGLOUGHLIN, DIRECTOR OF BIOTECHNOLOGY,
UNIVERSITY OF CALIFORNIA, DAVIS, 2000

The last chapter presented evidence that the great majority of the world's twenty-five most important food crops naturally hybridize with one or more wild relatives somewhere in the world. Is that list representative of domesticated species in general? Or is there something special about those twenty-five crops that make them different from most cultivated plants?

The quotations that start this chapter reveal contradicting perspectives on whether crop genes move into natural populations. Is spontaneous hybridization between crops and their wild relatives a "fact of nature" or a "rare exception"? Presently, there are two distinct views.

The first view corresponds to the McHughen quotation. For over a century (de Candolle 1886) most crop evolutionists have held the view that many, if not most, domesticated plants occur as part of a domesticate-weed-wild complex, defined by Anderson (1952) as "a compound of crops, accompanying weeds, and wild related species, mutually influencing each other by means of introgression." Theoretically, hybridization occurs sporadically in such complexes but frequently enough to allow regular exchange of alleles between their members (see fig. 8.1; see also de Wet and Harlan 1975, Jarvis and Hodgkin

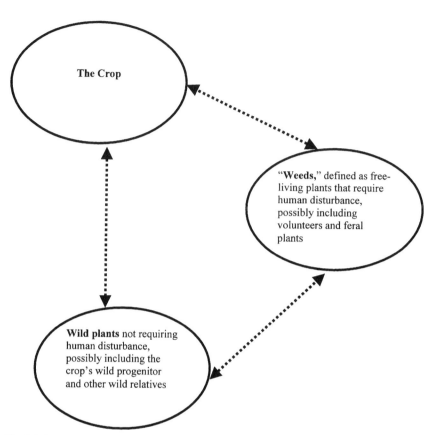

Fig. 8.1 Crop-weed-wild complex. Dashed arrows signify sporadic hybridization, providing opportunities for introgression. The amount of hybridization and introgression is idiosyncratic, varying with the specific populations involved.

1999, Ladizinsky 1998, Small 1984, Van Raamsdonk and Van der Maesen, 1996). These are folks who spend time in the field visiting regions where crops and their wild relatives often grow side by side. Their conclusion is largely based on the observation that at least a few plants sharing both domesticated and wild characters are typically found whenever crop plantations are growing in close proximity to their wild relatives. These plants are presumed to be hybrids.

The second view corresponds to the quotation from McGloughlin. Crop breeders who have focused on the difficulties of making crosses between crops and their wild relatives generally see the world from a perspective different from that of crop evolutionists. Prior to the advent of genetic engineering, deliberate introgression from a distant wild relative was often the only solution to delivering a badly needed trait to a crop. Executing such "wide crosses" is typically much more difficult than conducting crosses between different varieties of the same crop, sometimes requiring laboratory-based technologies such as embryo rescue, in vitro pollination, and somatic cell hybridization (Gill 1989, Ladizinsky 1989). Thus, when considering the likelihood of spontaneous hybridization, experts on such "wide hybridization" often emphasize the reproductive barriers between crops and their wild relatives (e.g., Ladizinsky 1985).

In this chapter I present relevant data from a wide variety of cultivated plants to address this controversy. Those data tend to support, at least on a global scale, the first view. Then I present data at the national scale: what is known regarding whether a country's important cultivated plants naturally hybridize with local wild plants. Again, these data also support the generality of cultivated-to-wild gene flow at that scale. I continue with a discussion of why evolution under domestication should not necessarily erect strong barriers to gene flow between domesticated plants and their wild forebears or other wild relatives. I conclude with the observation that some level of spontaneous gene flow and introgression between cultivated plants and their wild relations appears to be the rule.

The first line of evidence for the generality of domesticated-to-wild hybridization is the number and variety of species that engage in such hybridization. There are dozens of reports of natural mating between domesticated species and their wild relatives beyond those for the twenty-five crops detailed in chapter 7. Table 8.1 presents a list of cultigens for which there is evidence of natural hybridization with wild relatives under field conditions. This list combines the crops treated in detail in chapter 7 with additional examples I encountered while writing this book. The list is not exhaustive, and I have not attempted to obtain a thorough list of all examples because spontaneous hybridization among domesticated and wild plants is a difficult topic to search for in bibliographic databases. Nonetheless, table 8.1 lists eighty-one cultigens that apparently naturally mate with wild relatives. For each entry I have listed one representative supporting citation. In considerably more than half of the cases (48) support for hybridity comes from more evidence than just the presence of intermediate plants where the crop and wild relative are sympatric.

The crops evaluated in chapter 7 were diverse. Those in table 8.1 are even more diverse. The list includes plants used for all types of consumables—fruit,

Table 8.1 Evidence for Natural Hybridization between Domesticated Plants and Their Wild Relatives

Cultigen	Scientific Name	Evidence[a] for Hybridization	Representative Citation
Alfalfa	*Medicago sativa*	+	Jenczewski et al. 1999
Almond	*Amygdalus communis*	m	Ladizinsky 1998
Amaranth, grain	*Amaranthus spp.*	m	Sauer 1995
Apple	*Malus × domestica*	+	Dickson et al. 1991
Apricot	*Prunus armeniaca*	m	Mehlenbacker et al. 1990
Avocado	*Persea americana*	+	Ellstrand et al. 1986
Banana	*Musa acuminata*	+	Simmonds 1962
Barley	*Hordeum vulgare*	m	See chapter 7
Bean, common	*Phaseolus vulgaris*	+	See chapter 7
Beet, sugar	*Beta vulgaris*	+	See chapter 6
Bentgrass, common	*Agrostis capillaris*	m	DeVries et al. 1992
Bentgrass, creeping	*Agrostis stolonifera*	+	Wipff and Fricker 2001
Blueberry	*Vaccinium* spp.	m	Rosskopf 1999
Cacao	*Theobroma cacao*	+	R. Whitkus, personal comm.
Cane, sugar	*Saccharum officinarum*	+	See chapter 7
Canola (see rape)			
Cassava	*Manihot esculenta*	+	See chapter 7
Carrot	*Daucus carota*	m	Wijnheimer et al. 1989
Chayote	*Sechium edule*	m	Newstrom 1991
Cherry	*Prunus cerasus*	m	Wójcicki and Marhold 1993
Cocona	*Solanum sessiliflorum*	+	Salick 1992
Coffee, arabica	*Coffea arabica*[b]	+	See chapter 7
Coffee, robusta	*Coffea canephora*[b]	m	See chapter 7
Corn (see maize)			
Cotton	*Gossypium barbadense*	+	See chapter 7
Cotton	*Gossypium hirsutum*	+	See chapter 7
Cowpea	*Vigna unguiculata unguiculata*	m	See chapter 7
Daffodil	*Narcissus pseudonarcissus*	m	DeVries et al. 1992
Elm, Siberian	*Ulmus pumila*	+	Cogolludo-Augustín et al. 2000
Fescue, tall	*Festuca pratensis*	+	Borrill 1975
Fig	*Ficus carica*	m	Zohary 1995
Flax	*Linum usitatissimum*	m	Durant 1976
Gourd	*Cucurbia pepo*	+	Kirkpatrick and Wilson 1988
Grape	*Vitis vinifera*[b]	+	See chapter 7
Guava	*Psidium guajava*	m	Landrum et al. 1995
Hemp	*Cannabis sativus*	m	Small 1984
Hibiscus, Chinese	*Hibiscus rosa-sinensis*	m	Wagner et al. 1990
Hops	*Humulus lupulus*	m	Small 1984
Juniper	*Juniperus chinensis*	+	Ashworth et al. 1999
Lettuce	*Lactuca sativa*	+	Anderson 1949
Maize	*Zea mays* ssp. *mays*	+	See chapter 7
Millet, finger	*Eleusine coracana coracana*	m	See chapter 7
Millet, foxtail	*Setaria italica*	+	Till-Bouttraud et al.1992

Table 8.1 Continued

Cultigen	Scientific Name	Evidence[a] for Hybrid- ization	Representative Citation
Millet, pearl	*Pennisetum glaucum*	+	See chapter 7
Mint	*Mentha spicata*	m	Harley 1975
Mushroom, button	*Agaricus bisporus*	+	Xu et al. 1997
Oats	*Avena sativa*	+	See chapter 7
Olive	*Olea europaea europaea*	m	See chapter 7
Palm, coconut	*Cocos nucifera*	m	See chapter 7
Palm, date	*Phoenix dactylifera*	m	Wrigley 1995b
Palm, oil	*Elaeis guineensis*	m	See chapter 7
Pea	*Pisum sativum*	m	Davies 1995
Pear	*Pyrus pyrifolia*	m	Bell 1990
Pecan	*Carya illinoiensis*	m	Thompson and Grauke 1990
Pepper, black	*Piper nigrum*	m	Ravindran et al. 2000
Pigeonpea	*Cajanus cajan*	m	Van der Maesen 1986
Poplar, hybrid	*Populus trichocarpa × P. deltoides*	+	DiFazio et al. 1999
Potato	*Solanum ajanhuiri*	m	Johns et al. 1987
Potato	*Solanum stenotomum*	+	Rabinowitz et al. 1990
Potato	*Solanum tuberosum*	+	See chapter 7
Pyrethrum	*Chrysanthemum cinerariifolium*	m	Contant 1976
Quinoa	*Chenopodium quinoa*	+	Wilson and Manhart 1993
Radish	*Raphanus sativus*	+	Panetsos and Baker 1967
Rape, swede	*Brassica napus*	+	See chapter 7
Rape, turnip	*Brassica campestris*	+	See chapter 7
Raspberry	*Rubus idaeus*	+	Luby and McNichol 1995
Rhododendron, catawba	*Rhododendron catawbiense*	+	Milne and Abbott 2000
Rice	*Oryza glaberrima*	+	See chapter 7
Rice	*Oryza sativa*	+	See chapter 7
Rubber	*Hevea brasiliensis*	m	Wycherley 1995
Rye	*Secale cereale*	+	See chapter 7
Ryegrass	*Lolium perenne*	+	Lewis 1975
Safflower	*Carthamus tinctorius*	m	Knowles & Ashri 1995
Salsify	*Tragopogon porrifolius*	+	Novak et al. 1991
Sorghum	*Sorghum bicolor bicolor*	+	See chapter 7
Soybean	*Glycine max*	+	See chapter 7
Squash	*Cucurbia pepo*	+	Kirkpatrick and Wilson 1988
Strawberry	*Fragaria × ananassa*	+	Westman et al. 2000
Sunflower	*Helianthus annuus*	+	See chapter 7
Walnut	*Juglans regia*	+	Hussendoerfer 1999
Watermelon	*Citrullus lanatus*	+	Singh 1978
Wheat, bread	*Triticum aestivum*	+	See chapter 7
Wheat, durum	*Triticum turgidum durum*	+	See chapter 7

[a]m, morphological intermediacy only; +, more substantial evidence for hybridization.
[b]Other taxa account for a small portion of the world production of the crop.

vegetable, grain, spice, drink—and purposes other than food, such as fodder (e.g., alfalfa), industrial chemicals (e.g., pyrethrum), ornamentals (e.g., daffodil), turf grasses (e.g., creeping bentgrass), and so on. A few, such as hybrid poplar, might be considered "cultivated" but not strictly "domesticated." The list even includes a crop that isn't strictly a plant (button mushroom). A quick perusal of the table reveals that a cultivated plant need not be economically important or widely planted to engage in hybridization with wild relatives.

Of course, certain studies have found no evidence for cultivated-to-wild gene flow between closely related cultivated-wild pairs. Some of these studies are listed in table 4.1. Also, a descriptive genetic study of two bush mango *(Irvingia)* species using RAPDs as genetic markers found "no evidence of hybridization, even between individuals from areas of sympatry where hybridization was considered probable" (Lowe et al. 2000). But such reports are very rare. The paucity of such reports may be due in part to the reluctance of scientists and journals to publish "null results." And certainly we do not expect natural hybridization to occur among distantly related species.

Nonetheless, it is clear that many domesticated plants, if not the great majority, naturally mate with at least one wild relative somewhere in the world. But are these highly localized cases, or are they geographically widespread phenomena? Some scientists consider hybridization to be highly local, believing that "most crops have no interbreeding relatives in much of the world" (Gressel 1999). If most cultivated-wild hybridization occurs in a tiny part of the crop's range, then for any given region we would expect to find only a small percentage of crops naturally hybridizing with local wild species. But if such hybridization is widespread, then we would expect to find much more than a small percentage.

A few databases have been developed to examine the prevalence of local mating of domesticates with wild relatives. Concerns about hybridization as an avenue for the escape of crop transgenes has stimulated the creation of reviews addressing whether a country's important cultivated plants are likely to mate with wild taxa in the local flora. I am aware of such reviews for four different European countries. In the United Kingdom, Raybould and Gray (1993) examined the literature on thirty-one local domesticated species (excluding ornamentals) that have been genetically engineered and are most likely to be commercially released. They found that about one-third of that sample are known to spontaneously hybridize with one or more elements of the British flora. For the Netherlands, the analysis was based on both the literature and herbarium materials, and woody species were specifically excluded. About one-quarter of the forty-two important cultivated species considered reportedly naturally hybridize with one or more elements of the Dutch flora (deVries et al. 1992). Nurminiemi and Rognli (1993) examined a diverse list of 185 species that are deliberately cultivated within Norway, including forages, ornamentals, herbs, spices, and forest trees. Over 40 percent (77) of these cultigens readily hybridize with taxa in the Norwegian flora. The fourth example is Switzerland. Ammann et al. (1996) considered twenty-two important Swiss crops; over half (13) spontaneously hybridize with wild plants in the flora of that country. Thus, in this sample the fraction of domesticated species with local wild mates ranges from about 25 percent to 59 percent.

These proportions are surprisingly high. After all, the vast majority of Europe's cultivated species were domesticated elsewhere, and it is in those "centers of domestication" that we would expect the greatest concentration of close wild relatives to occur (Vavilov 1929). Also, the four countries mentioned are small and are in a part of the world whose natural flora is relatively species-poor. We might expect larger countries, those located in a "center of domestication," or those in the species-rich tropics to have an even higher frequency of domesticated species that mate with local wild relatives than the examples given. Unfortunately, reviews for such countries are apparently not available. Given that the United Kingdom, the Netherlands, Norway, and Switzerland represent countries least likely to hold wild relatives, perhaps only in the very smallest countries would domesticated-to-wild gene flow not occur at all.

What about the United States? It has been suggested that "most major agricultural crops lack free-living relatives (and therefore the opportunity to hybridize) in the United States because they originated in other areas of the world" (U.S. Congress, Office of Technology Assessment [OTA] 1993). I am not aware of any report that evaluates the opportunities for mating between the major crops of the United States and their local wild relatives. Therefore, I chose the twenty most important crops (in terms of area planted) in the United States. I used USDA reports (USDA 2001a, 2001b, 2001c, 2001d) to obtain the most recent (2000) estimates of area harvested for each crop. All but three of the twenty most important American crops of 2000 are a subset of those twenty-five reviewed in chapter 7, making evaluation relatively straightforward. That evaluation is summarized in table 8.2.

Table 8.2 Spontaneous Hybridization between the Most Important Crops of the United States and Their Wild Relatives

Rank	Crop	Area Planted[a] (10^6 Hectares)	Evidence[b] for Hybridization in U.S.	Evidence[b] for Hybridization Anywhere	Representative Citation for U.S. Hybridization
1	Maize (including sweet corn)	32.5	None	+	n/a
2	Soybean	30.1	None	+	n/a
3	Wheat	25.3	+	+	Seefeldt et al. 1998
4	Alfalfa	9.3	m	+	Great Plains Flora Association 1986
5	Cotton	6.3	m	+	Fryxell 1979
6	Sorghum	4.3	+	+	Arriola and Ellstrand 1996
7	Barley	2.4	None	m	n/a
8	Oats	1.8	m	+	Baum 1977
9	Rice	1.2	+	+	Langevin et al. 1990
10	Sunflower	1.1	+	+	Linder et al. 1998
11	Beans, dry and snap	0.8	None	+	n/a
12	Rapeseed (incl. canola)	0.6	+	+	Manasse 1992
13	Beets, sugar and table	0.6	+	+	Bartsch and Ellstrand 1999
14	Groundnut (peanut)	0.5	None	None	n/a
15	Potato	0.5	None	+	n/a
16	Rye	0.5	+	+	Sun and Corke 1992
17	Sugarcane	0.4	None	+	n/a
18	Grape	0.4	+	+	Olmo and Koyama 1980
19	Orange	0.3	None	None	n/a
20	Flaxseed	0.2	None	m	n/a

[a]Estimated area of production in U.S. for 2000 (USDA 2001a, 2001b, 2001c, 2001d).
[b]m, morphological intermediacy only; +, more substantial evidence for hybridization.

Spontaneous hybridization between the most important crops of the United States and their wild relatives is surprisingly general and does not agree with the expectation expressed by OTA and quoted in the last paragraph. Supporting evidence for natural hybridization within the United States is available for more than half of the crops (11, or 55 percent). Also, all but two of the twenty—groundnut (peanut) and orange—are known to naturally hybridize with wild relatives somewhere in the world. Clearly, the fact that most U.S. crops originated elsewhere does not impede the opportunity for regional hybridization with wild relatives. The reason is that many of the major crops are closely related to weeds of agriculture, which have migrated far beyond their place of origin with unintended human assistance. Compatible pairs of non-native American crops and weeds include wheat and goatgrass, sorghum and johnsongrass, and cultivated rice and red rice; hybridization among these pairs is discussed in chapter 7. Exotic weeds also account for many of the compatible crop relatives reported in the four European reviews discussed above. Clearly, a substantial fraction (although not necessarily a majority) of a country's crops naturally mate with wild plants within its borders.

Natural hybridization with at least one wild relative is the rule for domesticated plants. Furthermore, such hybridization apparently is a common feature of the major crops in any given region. This makes sense. The process of domestication starts with wild plants, which may be capable of hybridizing with other closely related species, as noted in chapter 3. Indeed, domesticated plants may still be capable of occasional spontaneous hybridization with both their progenitors and other species. Is there any reason to assume that the process of domestication and subsequent genetic improvement should erect substantial reproductive barriers between crops and their wild relatives?

Yes and no. Plant geneticists have two very different views on how domestication should impact reproductive isolation between the nascent domesticate and its progenitor. One school contends that domestication must include the evolution of reproductive isolation (Ladizinsky 1985). The key focus of this contention is that, as explained in chapter 5, gene flow can counterbalance selection and prevent progressive evolutionary change. That is, without some isolation, domestication cannot be successful, because the gains of the human selectors will be lost each generation when their selected plants mate with their wild parents. Therefore, successful selection for traits desirable to humans must go hand in hand with isolation from progenitors.

The other school focuses on the fact that in traditional farming systems, managed populations are so small that additions of genetic variation are helpful for counteracting the negative effects of inbreeding and drift as well as for introducing a substrate for further gains under selection (e.g., Jarvis and Hodgkin 1999). In such systems occasional introgression from wild plants can provide that additional variation. In this second case it is assumed that farmers select for, rather than against, plants that hybridize with wild relatives.

Although it is clear that domestication and subsequent genetic improvement involve the selection of desirable traits from a variable population, the specifics of domestication are largely elusive. These specifics comprise the questions that the evolutionary geneticist loves to ask: What was the average size of the original populations under selection for domestication? How many generations were under selection? How intense was the selection pressure? How much variation was contributed to the domesticate by other populations via gene flow?

These questions may prove difficult or impossible to answer for most domesticated plants. The domestication of many, perhaps most, crops occurred thousands of years ago (Smartt and Simmonds 1995). Also, human manipulation of plant genetics doesn't really start and stop with the domestication process. In many cases, if not the majority, domestication was probably preceded by a substantial period of human cultivation that probably influenced plant evolution (e.g., Anderson 1997). And, of course, genetic improvement doesn't stop when domestication is complete but is an ongoing process. Modern improvement continues both through selection of novel types and by means of sophisticated tools such as quantitative genetics, marker-assisted selection, wide crosses accomplished by embryo rescue, the generation of mutations through exposure to radiation or mutagenic chemicals, and the introduction of novel genes from unrelated organisms via genetic engineering. Any of these processes, including those occurring before or after domestication, might result in genetic changes that could increase or decrease barriers to reproductive isolation. Nonetheless, given that we have a crude knowledge of the processes of domestication and subsequent genetic improvement, we can ask whether these processes result in any obvious changes in reproductive isolation while recognizing that the answers are quite speculative and tentative.

What do we know about how domestication has altered mating patterns between domesticated plants and their wild progenitors? Spontaneous hybridization depends on spatial proximity, local availability of a pollen vector, simultaneous flowering, and compatibility. Let's consider each of these factors individually so as to ask whether human intervention in a cultivated plant's ecology and genetics has enhanced reproductive isolation between the domesticate and its wild relatives.

Spatial Proximity

Does domestication move useful plants far enough away from their wild relatives to prevent hybridization? Not necessarily. We wouldn't expect that crops would have evolved ecological requirements much different from their wild relatives, especially their progenitors. At the very least, for crops that were domesticated in ancient times, we would expect to find most crops growing in close proximity to and hybridizing with wild relatives near the region of their origin. Indeed, that is the case for many of the crops reviewed in chapter 7. Both pearl millet and finger millet are examples, growing in sympatry with wild relatives in Africa. Spontaneous intermediates, plants of probable hybrid heritage, are commonly seen in those mixed stands. Of course, as a crop becomes increasingly successful, it is often introduced far from its point of origin. For example, although tomatoes are an integral part of Italian cuisine, they were domesticated in the New World. No wild tomato relative yet grows as part of the natural Italian flora, and in that region spontaneous hybridization cannot occur.

But as a crop's range expands, so might the range of some of its wild relatives. Many crops have weedy counterparts that have evolved to live in cultivated or disturbed regions (de Wet and Harlan 1975). Even if these species were not the crop's progenitor, they still may have retained some ability to mate with the crop. Picked up on shoes and on tires, mixed with transported grain, blown by the wind, or sometimes intentionally introduced by humans (e.g., johnsongrass in the United States), these weeds often spread far beyond their homeland. Such immigrants often flourish in the same habitats as their crop relatives, becoming important weeds in the cultivated fields of their progenitors (goatgrass in wheat, johnsongrass in sorghum, etc.).

Chapter 5 discusses the fact that weeds related to crops may evolve to mimic their crop cousins. Sometimes hybridization and introgression may play a role in the evolution of crop mimics. Europe's weed beet, discussed in chapters 1 and 6, is an excellent example. It is curious that more wild relatives do not evolve to become crop mimics.

Furthermore, an introduced crop may have wild relatives waiting on distant shores. For example, domesticated grape and domesticated apple, both native to the Old World, have cross-compatible relatives in the New World; spontaneous hybridization followed their introduction (Smartt and Simmonds 1995). Clearly, it is unlikely that most crops will be spatially isolated from all of their wild relatives throughout their entire range of cultivation.

Pollen Vector

Does domestication alter the pollen vector? We wouldn't expect crops to evolve a pollination syndrome radically different from that of their wild relations. Generally, the pollen vector of a crop corresponds closely to that of close wild relatives. For example, interplant pollination of almost all species of the grass family, domesticated or not, is accomplished by the wind.

Furthermore, adequate pollinators may be waiting for a crop far beyond its place of origin. Wind and certain other pollen vectors are hardly localized to a narrow region. For example, even though the honeybee is native to the Old World, it has been introduced into the New World where it has spread widely (Buchmann and Nabhan 1996, Camazine and Morse 1988). Alternatively, native pollinators may adequately service an introduced crop.

Domestication may, however, involve a shift in terms of the need for a pollen vector. The domestication of plants often involves an evolutionary shift from obligate cross-pollination to some degree of self-fertility (Rick 1987). Examples include sunflower, tomato, and strawberry. Self-fertility may be selected for both as a mechanism that assures the production of a crop in the absence of pollinators and as a mechanism that increases the probability that a crop will "breed true" for desired traits (Allard 1999). But almost all plant species with high levels of self-fertilization still outcross to some extent (Lord 1981). And highly selfing plants are still capable of pollen dispersal over many meters (e.g., barley; Wagner and Allard 1991). We would not expect a shift from obligate outcrossing to facultative selfing to act as an absolute barrier to hybridization. Therefore, in general, the pollen vectors used by a crop's wild relatives should be available to effect crop-to-wild and wild-to-crop pollen transfer.

Simultaneous Flowering

Differences in flowering time are probably the most effective crossing barrier to plants that are otherwise compatible (Levin 1978). To illustrate, if species A flowers only in May and the cross-compatible species B flowers only in September, opportunities for spontaneous mating are zero. Is there any reason to expect that the domestication process or cultivation practices would result in a crop flowering at a time different from its progenitor or other wild relatives? While timing of flowering might not be a trait of great importance during the initial process of domestication, it is often one of great importance in the development of modern varieties. Agronomic crops that flower and bear fruit simultaneously are more easily cared for and can be harvested simultaneously (Marshall 1977). Modern maize cultivars are sexually active for only a couple of weeks; closely related wild teosinte taxa can flower for months. Refinement of varieties often involves genetic fine-tuning to create "early" or "late" cultivars. For example, an array of different Washington navel orange cultivars now make it possible to obtain the fruit fresh for most of the year (T. Kahn, personal communication).

Those differences in timing can have an impact on mating with wild relatives. Langevin et al. (1990) measured rates of spontaneous hybridization between the weed red rice and each of six cultivated rice varieties. The cultivar with the highest hybridization rate was the one with the greatest overlap in flowering time with the weed. Still, although some cultivars have flowering seasons different from those of their wild relatives, there is no reason to imagine that all or even most cultivars will have flowering times different from those of their wild progenitors.

Compatibility

As stated earlier in this book, plant breeders often find it a challenge to execute successful crosses between crops and certain wild relatives. Sometimes it is necessary for breeders to undertake "heroic" efforts to create hybrids, such as the rescue and culture of hybrid embryos that would otherwise be aborted (Sharma et al. 1996). Indeed, certain crops present "classical" cases for the evolution of reproductive isolation (Grant 1981), especially with regard to the evolution of cross-incompatibility. Such isolation suggests that domestication may include selection for reproductive isolation.

But if domestication involves intentional human-mediated selection of true breeding plants, the resulting reproductive isolation would be for reducing the production of hybrid seed by the domesticate. That is, the process of domestication will tend to select for isolating barriers that prevent the domesticate from being the female parent of hybrids. The process should not necessarily provide any selection pressure in the opposite direction, that is, hybridization when the wild plant is the female parent. Even a proponent of the view that domestication should erect isolating mechanisms wrote, "Gene flow, if it exists, is apparently more effective in the direction from the cultivated to the wild populations" (Ladizinsky 1985).

Therefore, if gene flow is a problem for plants undergoing domestication, we might expect the evolution of unilateral barriers to crossing. If hybrid production is somewhat rare because individuals of the crop are planted closer to one another than to wild plants that are the source of genetic pollution, additional isolation barriers need not be terribly strong ones. Perhaps the evolution of self-compatibility alone (Rick 1987) would suffice to reduce gene flow to tolerable levels. Furthermore, if domesticating farmers favor a small amount of novel variation in their crop (Jarvis and Hodgkin 1999), they may exert balancing selection that optimizes, rather than minimizes, gene flow. In any case, while it is clear that the processes of domestication and cultivation generally act to reduce gene flow between cultivated plants and their wild relatives, those processes should not necessarily act to minimize that gene flow.

In conclusion, the domestication process will probably always involve the erection of some barriers that reduce the probability that wild relatives will contribute to the gene pool of the crop. But there is no reason to assume that the erection of these barriers must be bilateral or absolute. Domestication is an evolutionary process under artificial selection. The accumulation of barriers to reproductive isolation is not necessarily rapid. Furthermore, as we have seen in chapter 3, plant populations that are otherwise considered to be "good" species may occasionally hybridize under the right circumstances. Is there any reason plants that have diverged under artificial selection should be different from those that diverged under natural selection? The patterns are in fact very similar. Just like naturally evolving species, occasional spontaneous hybridization with close relatives is a general feature of domesticated plants.

And just as with natural plant species, the amount of hybridization and introgression that occurs between a given crop and a wild relative is going to be idiosyncratic over time, space, and the species involved. In a few cases (peanut, chickpea, and bush mango), spontaneous hybridization may not occur at all. In others, hybridization occurs with some closely related taxa (e.g., *Zea mays* ssp. *mays* and *Z. mays* ssp. *mexicana*) and not others (e.g., *Zea mays* ssp. *mays* and *Z. perennis*). In certain cases, hybridization appears to be largely restricted to first-generation hybrids (e.g., apparent wheat × goatgrass hybrids). In yet others, hybridization and introgression appear to occur to such an extent as to exemplify the "crop-weed-wild complexes" described by many plant evolutionists. Such complexes have certainly been shown to occur for beet in Europe (Viard et al. 2002), maize in Mexico (Blancas 2001), radish in California (Panetsos and Baker 1967), and soybean in Japan (Abe et al. 1999), to name a few. Indeed, such complexes may occur for the majority of cultigens. The thorough population-genetic-level studies necessary to test whether substantial introgression of crop alleles into the wild has occurred are still lacking for the majority of the cases listed in table 8.1. They offer fertile opportunities for future research.

And even more important is whether that introgression has had significant evolutionary or applied impacts on the wild populations that have been the recipients of crop alleles. What is known of those impacts is the topic of chapter 9.

Part III / Dangerous Liaisons?

Some Impacts of Gene Flow of Domesticated Alleles into Wild Populations

Go with the flow.   —ANONYMOUS

Here begins the third and final part of this book. The first part of the book introduced hybridization and its consequences. The second part revealed that natural hybridization with wild plants is a general feature of domesticated plants. The final part of this book addresses the impacts of those liaisons, especially their significance for the "special" case of transgenic crops.

Spontaneous hybridization between wild plants and domesticated plants is frequent enough to have been documented for dozens of species. But what does that mean for the wild populations? Under the appropriate circumstances, the gene flow that a population receives can play an important role in its evolution (summarized in tables 4.2 and 4.3). Gene flow can alter population-genetic diversity and mean fitness. It can lead to the evolution of new species or can break down reproductive barriers between species. Or, under other circumstances, gene flow can have no effects at all.

Occasionally the flow of domesticated alleles into natural populations has had more significance than as a mere topic for discussion by academics. Gene flow can sometimes lead to the evolution of new weeds (or invasives) or the evolution of greater aggressiveness in existing weeds. But gene flow can also be an agent for extinction, either by reducing the fitness of hybridizing plants or by permitting the assimilation of one species into the gene pool of another. Chapter 4 laid out the theoretical consequences of domesticated gene flow into wild populations. This chapter examines the actual consequences of that gene flow: whether gene flow from domesticated species to their wild relatives has resulted in any of the predicted impacts.

What is the evidence that gene flow by hybridization with domesticated species has had a significant impact on wild populations? Some of the case studies in chapters 6 and 7 present specific instances in which gene flow from crops has actually had an evolutionary impact on wild populations. For example, in several cases it appears that introgression is rampant. The case studies in chapters 6 and 7 also make it clear that spontaneous hybridization has been economically and environmentally significant in terms of weed evolution and putting wild taxa in jeopardy. I consider each of these possible gene-flow impacts separately and address the question of whether the evidence supports the predicted impacts of domesticated gene flow into wild populations.

Changes in Genetic Diversity in Natural Populations?

Gene flow has the potential to significantly increase or decrease a population's genetic diversity (table 4.2). It is widely believed that if gene flow occurs among wild and cultivated plants, it should have important consequences for the amount of diversity in both kinds of populations (e.g., Jarvis and Hodgkin 1999). And indeed, the descriptions of putative domesticated-wild hybrid swarms often describe a striking amount of morphological variation compared with presumably genetically pure populations of both parents; examples include maize (Wilkes 1977) and finger millet (de Wet 1995a). Still, judgments about genetic diversity based on morphology are often anecdotal and qualitative, without genetic analysis or quantitative comparison.

Even experimental and descriptive genetic analyses of introgressed populations have generally focused on documenting the fact of hybridization between domesticates and wild plants rather than measuring its consequences. Genetic analysis of a relatively small number of appropriate plants (e.g., a few dozen or fewer) at a few loci is all that is necessary to provide strong evidence for spontaneous hybridization. But evidence for an evolutionary impact at the population level often requires more thorough analysis.

Studies that document spontaneous hybridization between domesticates and their wild relatives rarely ask whether and how gene flow is associated with a change in genetic diversity. Nonetheless, a handful of studies have directly addressed whether gene flow from a crop has altered genetic diversity in the wild population. Some are experiments to measure genetic changes in a natural population following a bout of gene flow from a related cultivar. Others are descriptive studies that quantitatively compare the genetic variation in wild populations that received gene flow from a crop with other control populations that apparently did not.

The experimental studies are informative regarding the spatial and temporal dynamics of individual alleles. Many studies have characterized genetic variation in natural populations that have been receiving gene flow from a related crop. But to my knowledge only two studies obtained allele frequencies of wild populations prior to receipt of gene flow and thus measured the precise evolutionary genetic impacts due to gene flow from a crop. The first (Whitton et al. 1997) characterized the frequency of two cultivar-specific RAPD markers in a natural population of sunflower *(Helianthus annuus)* in northern California before and after cultivated sunflower (the same species) was planted in an adjacent field for the first and only time. Those RAPD markers were initially absent in the wild population and occurred at an initial frequency of 0.5 in the cultivar. Introgression was measured by analysis of progeny collected from the wild population. Progeny were sampled the first year, immediately after natural hybridization had occurred (generation 1). Three meters from the crop margin, 42 percent of the progeny tested proved to be hybrids. Hybrids were also detected at two hundred meters and four hundred meters from the crop margin, but at a much lower rate (≤10 percent). Progeny were sampled from the same locations in the wild population again at generation 3 and generation 5. Both alleles persisted and spread through the population. The overall allele frequencies did not change over the generations, suggesting that these particular markers (and the alleles to which they are closely linked) are more or less neutral in the wild. But the spatial-specific frequencies changed such that the frequencies became more uniform throughout the population. The alleles diffused into the population, decreasing in frequency near the site of the crop margin and increasing at the two-hundred- and four-hundred-meter collection sites. Homogenization and persistence occurred even though first-generation cultivated-wild sunflower hybrids are somewhat less fit in the wild than their genetically pure wild sibs (Snow et al. 1998, Cummings et al. 1999).

In another experiment Snow et al. (2001) measured the impact of gene flow from cultivated radish *(Raphanus sativus)* to jointed charlock *(R. raphanistrum)* in four field sites in northern Michigan. First-generation interspecific hybrids were initially planted intermixed with the wild species in a 1:1 ratio. The genetic marker monitored was the cultivar's white flower color allele, which is dominant to the yellow allele specific to the wild species. Therefore, the initial allele frequencies were 1:3 white:yellow (in year zero). With random mating and neutral gene flow, at equilibrium, one would expect about 43 percent of the plants in these populations to be white-flowered. When they were revisited two and three years later, the experimental sites had about 7–22 percent white-flowered plants. The drop in allele frequency is not surprising, given that the initial hybrids set substantially fewer seeds than the pure *R. raphanistrum* (Snow et al. 2001). What is surprising is that the white-flowered phenotype showed a slight increase in three of the four populations from year two to year three. The combined data suggest that gene flow is not initially neutral but detrimental, because of lower relative fitness of the hybrids; however, it appears that once recombination has occurred, the allele (or those to which it is tightly linked) is advantageous in the wild. Because it is not possible to score heterozygotes from phenotype alone with a dominant marker, it is not clear that the allele frequency is actually increasing in these populations. Likewise, the standing population might not yet be in genetic equilibrium, because wild radish seeds can remain dormant in the soil over one or more years. Only repeated sampling of these sites will reveal the final fate of the crop allele. Both the sunflower and radish multigeneration experiments show the increase in genetic diversity in a wild population to be expected from a single pulse of crop gene flow (table 4.3).

What about the evolutionary impact of repeated unilateral crop gene flow into wild populations (table 4.3)? In this case we turn to descriptive studies that compare the genetic structure of wild populations that grow adjacent to cross-compatible cultivated plants with wild populations that presumably have always been isolated from cultivated relatives. Blancas (2001) genetically analyzed populations of the wild taxon teosinte *(Zea mays* ssp. *mexicana)* at eighteen allozyme loci. She then compared the genetic diversity of populations that were intermingled with cultivated maize *(Z. mays* ssp. *mays)* with those that were isolated from that taxon. The comparison was made after all morphologically intermediate individuals had been removed for separate analysis. Even after excluding those plants, the pattern of allozyme variation confirmed introgression of maize alleles into teosinte (see the more detailed discussion in chap. 7). Most important, those introgressed populations had more alleles per locus and more polymorphic loci than genetically pure isolated populations of *Z. mays* ssp. *mexicana*. This trend runs counter to the long-term predictions stated in chapter 4 but in this case is apparently explained by the fact that the source of the domesticated gene flow, maize land races, are considerably more polymorphic than the pure populations of the wild recipient (Blancas 2001). The common assumption (and observation) that domesticated plants are typically less polymorphic than their wild relations (see chap. 4; also Doebley 1989, Ladizinsky 1985) does not hold in this specific case.

Another comparative study involves cultivated gene flow into populations of wild beets. Wild sea beets of the Adriatic coastal plain of northeastern Italy grow near cultivated beet seed multiplication sites, sometimes only one thousand meters away (Bartsch et al. 1999). Bartsch et al. (1999) compared the population-genetic constitution of these wild beets with that of wild populations isolated from cultivated beets. They compared allele frequencies of eleven allozyme loci in the two samples. Alleles that were common in the cultigens but absent or quite rare in the isolated wild populations were common in the wild populations growing close to the seed multiplication regions. These data support the idea that these populations are introgressed with crop alleles, and complement prior studies using genetically based morphological markers (Bartsch and Schmidt 1997, Bartsch and Brand 1998). As a group, the wild populations receiving gene flow from the cultivated beets were generally about as variable as, or more variable than, those populations that were isolated from the crop for a number of standard genetic diversity parameters. Compared with the isolated populations, the introgressed populations had a substantially higher fraction of polymorphic loci and a substantially higher level of gene diversity. The differences in diversity are especially impressive given that the isolated populations embrace a substantial chunk of geography over coastal Europe and the introgressed populations are restricted to a tiny slice of the northern coast of the Adriatic Sea.

Again, the trend runs counter to the predicted consequences of long-term gene flow stated in chapter 4. A close inspection of the allele frequencies in these wild beet populations reveals some interesting patterns that explain this

apparent paradox. Typically, the allele frequencies of the introgressed populations fall between the one of beet cultigens (sugar beet or red beet/Swiss chard)
and that of the isolated wild populations. At some loci the introgressed populations tend toward sugar beet; in others they tend toward red beet/Swiss
chard (those two cultigens are too similar to distinguish). The pattern suggests
that the wild populations have absorbed alleles from at least two kinds of cultigens. The conclusion is not surprising, because all of the cultigens have been
grown in the area. The predictions in chapter 4 are based on the assumption
of several generations of gene flow from a single cultivar into a wild population, rather than multiple divergent sources of domesticated alleles.

In addition to the four studies detailed above, apparent changes in diversity due to gene flow can be deduced from population-genetic studies conducted for a variety of other reasons. Several more studies have measured both
the population-genetic structure of one or more wild populations that received gene flow from a cultivated relative as well as that of one or more wild
populations that had a history of isolation from related cultivars. I have used
those data to determine whether and how gene flow apparently altered diversity of the wild populations. I considered only studies where it is clear that hybridization and introgression have actually occurred; that is, the genetic analysis demonstrated that feral and wild plants are not simply intermixed. Those
studies, plus the four just discussed, are listed in table 9.1. With a single exception, the general trend is toward increased diversity in those wild populations
receiving gene flow from a cultivated relative. Many explanations could be offered for the trend, but caution should be exercised before making too much
out of this generalization. In the two experimental studies gene flow is so recent that the only change observed would necessarily be in the direction of increased diversity. Likewise, in some of the descriptive reports the only alleles
under scrutiny are those that were initially absent in the wild populations,
and as such their debut in the wild would add to increased polymorphism until they reach fixation. Finally, scientists who found few or no changes in allele frequencies might be reluctant to attempt to publish "null results."

Fortunately, Luby and McNichol (1995) did not experience that apprehen-
sion. Their research represents the exceptional entry in table 9.1. They measured the frequency of two crop-specific alleles in populations of Scottish wild
raspberry. Despite decades of opportunity for hybridization with domesticated plants, wild raspberry populations growing near the crop were hardly
different from those growing in continuous isolation. The very low frequency

Table 9.1 Genetic Diversity Changes in Introgressed Relative to Isolated Wild Populations

Cultigen Gene-Flow Source	Wild Gene-Flow Recipient	Nature of Study	Genetic Markers Compared	Relative Diversity in Introgressed Populations	Citation
Beet, *Beta vulgaris* ssp. *vulgaris*	*B. macrocarpa*	Descriptive	Allozymes	Greater[a]	Bartsch and Ellstrand 1999
Beet, *Beta vulgaris* ssp. *vulgaris*	*B. vulgaris* ssp. *maritima*	Descriptive	Allozymes	Greater[a]	Bartsch et al. 1999
Maize, *Zea mays* ssp. *mays*	*Z. mays* ssp. *mexicana*	Descriptive	Allozymes	Greater[a]	Blancas 2001
Radish, *Raphanus sativus*	*R. raphanistrum*	Descriptive	Cytogenetic, morphological	Greater	Panetsos and Baker 1967
Radish, *Raphanus sativus*	*R. raphanistrum*	Experimental	Flower color allele	Greater	Snow et al. 2001
Rapeseed, *Brassica napus*	*B.campestris*	Descriptive	AFLPs	Greater	Hansen et al. 2001
Raspberry, *Rubus idaeus*	Same species	Descriptive	Two morphogical alleles	Introgression detected but near zero for these markers	Luby and McNichol 1995
Rye, *Secale cereale*	*S. montanum*	Descriptive	Allozymes	Greater[a]	Sun and Corke 1992
Siberian elm, *Ulmus pumila*	*U. minor*	Descriptive	Allozymes	Greater	Cogolludo-Agustín et al. 2000
Sunflower, *Helianthus annuus*	Same species	Experimental	RAPDs	Greater	Whitton et al. 1997
Sunflower, *Helianthus annuus*	Same species	Descriptive	RAPDs	Greater[b]	Linder et al. 1998
Sunflower, *Helianthus annuus*	*H. petiolaris*	Descriptive	AFLPs	Greater[c]	Rieseberg et al. 1999

[a]For most parameters compared, little difference for the rest.
[b]For most loci examined, but a few crop-specific alleles are now more common than the wild alleles.
[c]For most loci examined, but some did not introgress at all.

of the spineless allele in these populations, and its absence in isolated populations, confirms that some gene flow has occurred, but that allele is so rare that it appears that the effect of gene flow has been essentially negligible in this species.

Nonetheless, the generalization from the data accumulated here is that gene flow from domesticates tends to result in an increase of genetic diversity in the wild. If we take this generalization at face value, two reasonable explanations for the trend are likely. The first is that the typical assumptions about the nature of alleles flowing from the crop are too simplified, for example, the assumption that at any given locus the domesticated gene-flow source is supplying the wild sink population with a single allele that eventually supplants multiple alleles in the sink population. The maize example discussed above illustrated that the individual crop populations may sometimes be more polymorphic than the recipient wild populations. The beet example illustrated that multiple crop cultivars that hold different alleles can act collectively as a gene-flow source that is more variable than the recipient wild population. Also, the replacement of one cultivar by another over time may act to transmit different alleles into natural populations, such that those populations may act to accumulate "heirloom" collections of crop alleles (Blancas 2001). The second explanation is that domesticated gene flow has been too little and/or too recent for the novel crop allele(s) to reach fixation in the wild. In that case, the addition of the novel allele may make a population that was originally fixed for a single allele more variable because of the introduction of a second allele. A third explanation is that while gene flow from a crop might often lead to uniformity in a wild population, it is possible that such cases send the wild population to extinction so rapidly (see below) that little time is available to observe and study that process. Only experimental work involving multigenerational study of the impacts of substantial unilateral gene flow from a monomorphic source crop population into a polymorphic sink population will determine whether this latter explanation is a reasonable one.

Changes in Fitness of Individuals in Natural Populations?

Chapter 4 details why domesticated alleles might affect the fitness of plants in the wild. At one extreme, certain crop alleles would likely prove very detrimental in the wild (e.g., those that result in reduced levels of compounds that defend the plant from certain animals that eat it). At the other extreme, other alleles (say, those for disease resistance) might prove beneficial. Two sets of data have been collected that are relevant to the fitness impact of domesticated alleles in natural populations. One set is the fitness of the early generations following hybridization, based on either the measurement of relative fitness of artificially created F_1s growing under field conditions (table 4.4) or observations that putative natural crop-weed hybrids sometimes have partial or complete sterility (Stace 1975b). In both cases the known or apparent first-generation hybrids generally have reduced fitness, but exceptions to this trend are not rare (examples in table 4.4; see also Arnold and Hodges 1995). When F_1s have substantially reduced fitness, backcrosses to the wild species often show a substantial recovery in fertility (e.g., Mikkelsen et al. 1996, Zemetra et al. 1998). To my knowledge only one study has measured the relative fitness of advanced-generation domesticated-wild hybrids in the field. The work was conducted on F_2 derivatives of *Brassica napus* × *B. campestris* (Hauser et al. 1998a). Interestingly, on average, these plants showed "hybrid breakdown," lower field fitness than both their first-generation hybrid parents and their wild grandparent. But they also showed considerable variation between families, with some approaching the fitness of the genetically pure wild plants. It is likely that fitness effects in these F_1s and other early hybrids, whether positive or negative, are largely due to interactions between the divergent coexisting genomes of hybrids rather than the effects of individual alleles (Arnold 1997, Grant 1981). Thus, the foregoing data are better suited for predicting whether introgression will be inhibited or augmented than for predicting long-term fitness changes in an introgressed population after several generations of gene flow and recombination.

The other set of relevant data regarding fitness changes are examples of lineages that have enjoyed adaptive evolution following hybridization. These hybrid descendants have enjoyed the evolution of higher fitness than their parental lineages, at least in certain environments. Examples are given in tables 9.2 and 9.3. In some cases the hybrid-derived lineages have achieved a taxonomic epithet (table 9.2). In other cases a new lineage has been identified and studied but not named (table 9.3). In both types of cases it appears that some aspect of hybridization, although not necessarily the impacts of individual alleles, has stimulated adaptive evolution.

In a few cases it is clear that the genetic basis of the adaptation is associated with certain introgressed alleles. For the European weed beets that evolved in

Table 9.2 New Taxa That Evolved Following Hybridization between Domesticated and Wild Plants

Derived Taxon	Cultigen Parent	Wild Parent	Site of Taxon's Origin	Evidence[a] of Hybridity	Representative Citation	How Stabilized?	Agricultural Weed or Invasive?
Glycine gracilis	G. max, soybean	G. soya	Northeast Asia	+	Abe et al. 1999	Largely selfing line	Not particularly weedy or invasive
Nasturtium sterile	N. officinale, watercress	N. micro-phyllum	Europe	m	Stace 1991	Clonal growth	Disturbed area weeds
Pennisetum sieberanum	P. glaucum, pearl millet	P. violaceum	Africa	+	Brunken et al. 1977	Largely selfing line	Agricultural weed
Solanum × edinense ssp. edinense	S. tuberosum ssp. tuberosum, potato	S. demissum	Scotland	+	Hawkes 1990	Clonal growth	Not particularly weedy or invasive
Solanum × edinense ssp. salamani	S. tuberosum ssp. andigena, potato	S. demissum	Mexico	+	Hawkes 1990	Clonal growth	Agricultural weed
Sorghum almum	S. bicolor, sorghum	S. propinquum	South America	+	Paterson et al. 1995	Allopolyploid	Agricultural weed
Tragopogon mirus	T. porrifolius, salsify	T. dubius	United States	+	Novak et al. 1991	Allopolyploid	Increasing in range and numbers
Triticum dicoccoides "Upper Jordan Valley" race	T. turgidum spp. durum, durum wheat	T. dicoccoides	Near East	+	Blumler 1998	Largely selfing line	Not particularly weedy or invasive

[a] m, morphological intermediacy only; +, more substantial evidence for hybridization.

Table 9.3 *Examples of Adaptive Evolution Following Hybridization between Cultivated and Wild Plants*

Cultigen Parent	Wild Parent	Site of New Lineage's Origin	Evidence[a] of Hybridity	Representative Citation	How Stabilized?	Evolutionary Niche Change to . . .
Beta vulgaris ssp. *vulgaris*, sugar beet	*B. v.* ssp. *maritima*	Europe	+	See chapter 6	Coalescent complex	Agricultural weed
Eleusine coracana, finger millet	*E. c.* ssp. *africana*	Africa	m	De Wet 1995a	Not clear	Agricultural weed
Manihot esculenta, cassava	*M. reptans*	South America	m	Nassar 1984	Introgression of alleles	Range extension
Oryza sativa, rice	*O. sativa* f. *spontanea*	India	+	Oka and Chang 1959	?	Mimetic weed
Raphanus sativus, radish	*R. raphanistrum*	California	+	Panetsos and Baker 1967	Coalescent complex	Agricultural weed/invasive
Rhododendron catawbiense, Catawba rhododendron	*R. ponticum*	United Kingdom	+	Milne and Abbott 2000	Introgression of alleles	Invasive
Secale cereale, rye	*S. montanum*	California	+	Sun and Corke 1992	Coalescent complex	Agricultural weed/invasive
Sorghum bicolor, grain sorghum	*S. halepense*	United States	+	De Wet and Harlan 1975	Introgression of alleles	Increased weediness

[a]m, morphological intermediacy only; +, more substantial evidence for hybridization.

the last few decades (table 9.2, chaps. 1 and 6), the morphology and physiology of their sugar beet ancestor combined with the bolting allele of the wild beet resulted in a self-seeding weed that closely resembles the crop and is well suited for life in cultivated beet fields. Another example is *Rhododendron ponticum,* an invasive species in the British Isles that has colonized areas much colder than predicted by its native range in Iberia. This wider ecological tolerance is best explained by its history of hybridization with the cold-tolerant *R. catawbiense,* introduced into Britain as an ornamental from North America (Milne and Abbott 2000). Apparently, alleles from the latter species account for the increased niche of *R. ponticum.*

In addition to a niche change generated by the introduction of one or more alleles, it has been hypothesized that the increase in genetic variation produced in a hybrid lineage could, in itself, suffice as a mechanism of adaptive success (Stebbins 1969). This argument falls into the category of a "group selection" explanation. But invasiveness is essentially a group trait, not one that can be measured from a single individual. Weedy radish and weedy rye, both hybrid lineages arising in California in the last century, are potential examples of this type, with considerably more genetic variation than their wild progenitors (Panetsos and Baker 1967, Sun and Corke 1992). Although these examples are compatible with the genetic variation hypothesis, rigorous experimental work with these systems would be a better test to determine the basis of their adaptive success.

In other cases the evolution of adaptive change may be a result of genetic phenomena other than introgressed alleles from the domesticate per se. The fixed hybridity afforded by allopolyploidy or asexual reproduction may result in fixed hybrid vigor (Grant 1981) and may explain the evolution of invasiveness in many hybrid derivatives, whether or not a domesticated plant is one of the parents (Ellstrand and Schierenbeck 2000). For example, the aggressiveness of *Sorghum almum,* a noxious weed that is an allotetraploid derivative of a hybrid of grain sorghum *S. bicolor* and its wild relative *S. propinquum,* could be explained by either the combination of crop and wild alleles or its hybrid vigor maintained by its chromosomal hybridity (Paterson et al. 1995). Regarding domesticated-wild hybrids that are largely seed-sterile but capable of vegetative spread (e.g., *Solanum* × *edinense*), Baker (1974) remarked, "Wasting no energy on seed production, their considerable vigor is channeled into vegetative reproduction." Also, hybridization may lead to the evolution of invasiveness through an opportunity for populations with a history of isolation and small population size to overcome the genetic load of detrimental alleles accumulated by drift (Ellstrand and Schierenbeck 2000). Indeed, there is no reason to assume that the evolution of adaptive change is necessarily explained by a single genetic mechanism, because these explanations are rarely mutually exclusive (Ellstrand and Schierenbeck 2000).

Hybridization between cultivated species and their wild relatives has sometimes served as a stimulus for adaptive evolution. But I am aware of only one example of crop gene flow that has led to the evolution of decreased fitness in wild populations. Natural hybridization with *japonica* varieties of rice in Taiwan is associated with decreasing seed set and pollen fertility in the endemic wild rice *Oryza rufipogon* ssp. *formosana*, which contributed to its extinction in the wild (Kiang et al. 1979, Oka 1992). Despite the many examples of increased extinction risk due to gene flow from cultivated relatives (table 9.4), I am not aware of data beyond the case of this Taiwanese wild rice that demonstrate that the introgression of domesticated alleles has resulted in decreased fitness at the population level. This lack of data does not mean that the negative fitness effects of crop gene introgression are rare. Rather, the lack of observed fitness failures may have more to do with the speed at which extinction by hybridization occurs (see below) and the fact that such comparative fitness studies are not simple, requiring a comparison of adequate-sized samples of plants from generations prior to and after the start of substantial gene-flow receipt. Comparing the fitness of endangered wild populations receiving gene flow from a cultivar with those without a history of hybridization is a high priority for understanding the mechanisms under which extinction by hybridization occurs. Indeed, experimental work of this type would also be helpful in elucidating the details of how hybridization has led to the cases of adaptive evolution following hybridization that were discussed earlier.

Changes in Reproductive Isolation?

Hybridization can have three different effects on the evolution of reproductive isolation: isolation barriers between the parents can be reinforced, they can be diminished, or the hybrid lineage can evolve barriers that isolate it from both parents. It is not clear whether gene flow from cultivated plants to wild plants has reinforced the isolation barriers between them. While isolation barriers between domesticated plants and their wild relatives are well

Table 9.4 Examples of Hybridization with a Cultivated Species That Endangers Wild Populations

Hybridizing Cultigen	Scientific Name	Wild Populations at Risk	Location	Representative Situation	Citation(s)
Alfalfa	*Medicago sativa*	*M. falcata*	Switzerland	"An endangered species, threatened genetically by introgression"	Rufener Al Mazyad and Ammann 1999
Cherry	*Prunus cerasus*	*P. fruticosa*	Czech Republic and Slovakia	"Hybridization appears to be a major threat to" *P. fruticosa* in these areas	Wójcicki and Marhold 1993
Coconut	*Cocos nucifera*	wild *C. nucifera*	Indian and Pacific Oceans	No truly wild coconuts may be left	Harries 1995
Cotton	*Gossypium hirsutum*	*G. darwinii*	Galapagos Islands	At risk of extinction by hybridization	Fryxell 1979, Wendel and Percy 1990
Elm, Siberian	*Ulmus pumila*	*U. minor*	Spain	"Extensive hybridization" and disease "have endangered . . . the native elm"	Cogolludo-Agustín et al. 2000
Palm, date	*Phoenix dactylifera*	*P. canariensis*	Canary Islands	"Main threat seems to be hybridization with *P. dactylifera*"	Morici 1998
Rice	*Oryza sativa*	*O. nivara*	East Asia	Typical (nonintrogressed) individuals are now hard to find	Chang 1995
Rice	*Oryza sativa*	*O. rufipogon* ssp. *formosana*	Taiwan	Extinct or nearly extinct in the wild	Kiang et al. 1979, Oka 1992
Strawberry	*Fragaria* × *ananassa*	*F. virginiana*	United States	"It is difficult to find 'pure' populations in many areas"	Rosskopf 1999
Walnut	*Juglans regia*	*J. hindsii*	California	"Cultivated walnut may hybridize the . . . Hinds walnut out of existence"	Ledig 1992

known (examples in Smartt and Simmonds 1995), the initial evolution of those barriers is assumed (and sometimes known) to be a byproduct of the domestication process. One line of evidence for the evolution of reinforced isolation barriers would be if wild populations in sympatry with a crop had stronger isolation barriers than populations that were geographically distant from plantations of the crop. This pattern is found for two wild subspecies of *Zea mays*. Wild populations of *Z. mays* ssp. *mexicana* often grow intermixed with maize *(Z. mays* ssp. *mays);* the two have a strong incompatibility barrier (Kermicle 1997). The wild *Z. mays* ssp. *parviglumis* grows far from areas of maize cultivation and crosses much more easily with the crop. Even these data are only suggestive of reinforced isolation barriers, and I am not aware of other data supporting such evolution in other systems. Beyond the obvious isolation barriers of chromosomal differences (examples in Grant 1981, Smartt and Simmonds 1995) and the evolution of increased self-fertility in crops (Rick 1987), the whole question of the evolution of isolation barriers between crops and their wild relatives appears to be poorly studied.

Examples of the dissolution of isolation barriers are few but better documented. The case of wild radish in California is perhaps the best example of a "coalescent complex," in which the identities of parental species have disappeared into what seems to be a randomly mating, highly polymorphic hybrid lineage (Panetsos and Baker 1967). Cultivated radish, *Raphanus sativus,* and jointed charlock, *R. raphanistrum,* behave as discrete species in Europe. There they have species-specific differences in floral, fruit, and root morphology (see chap. 3 for details). In some parts of Europe the two species may hybridize when in contact, leading to localized hybrid swarms (e.g., in the United Kingdom; Stace 1975b), but in other parts (e.g., in the Netherlands; deVries et al. 1992) they apparently do not hybridize. The situation in California is radically different. There, jointed charlock and radish have hybridized and introgressed into a single polymorphic coalescent complex that extends over hundreds of square kilometers of California's coastal plains and inland valleys (Panetsos and Baker 1967). Nowadays in California it is impossible to find individuals that can be clearly assigned to *R. sativus* or *R. raphanistrum* (N. Ellstrand, J. Nason, J. Clegg, and S. Hegde, unpublished data). For example, most individuals have a taproot indistinguishable from that of *R. raphanistrum* and fruits with some intermediacy between characters of the two parental species. In this case, in populations comprising thousands of hectares, plants of hybrid ancestry have swamped their progenitors to near extinction. Other examples of coalescent wild-crop hybrid complexes covering large areas include Europe's weed beets (chaps. 1 and 6) and California's weedy rye (chap. 7).

The largest number of the reported evolved changes in reproductive isolation following hybridization between wild plants and cultivated relatives are in the category of newly evolved taxa that are isolated from both parents (examples in table 9.2). The evolution of *Tragopogon mirus* is not only a relevant example but also a "classical" example of rapid speciation by the evolution of polyploidy following hybridization (Novak et al. 1991). *T. porrifolius* (2n = 12), salsify, is a root crop native to Europe. It hybridizes readily with the wild *T. dubius* (2n = 12). *T. mirus* (2n = 4× = 24) is their allotetraploid derivative and is incompatible with both parental taxa, creating sterile triploid progeny when crossed to them. Analysis with an array of genetically based markers has demonstrated its hybrid origin and reveals that different lineages of *T. mirus* evolved from *T. porrifolius* × *T. dubius* hybrids, not once but multiple times, in the western United States (Soltis and Soltis 2000). Reproductive isolation of hybrid derivatives can also be maintained through clonal reproduction or a predominantly selfing breeding system (examples in table 9.2). Clearly, hybridization between crops and wild relatives can have important evolutionary consequences.

Evolution of Increased Invasiveness or Weediness?

Hybridization between crops and wild relatives can have important consequences for those who manage plant populations, whether of useful species or endangered ones. One concern to managers of both kinds of populations is the presence of unwanted aggressive species. In the agricultural context these are referred to as "weeds"; in natural ecosystems, as "invasives." De Wet and Harlan (1975) speculate that "seed crops hybridize with their ancestral races to produce weedy derivatives wherever wild and cultivated kinds are sympatric." Tables 9.2 and 9.3 include over a dozen well-established cases in which hybridization between a cultivated plant and a wild relative has preceded the evolution of increased aggressiveness of the wild taxon. The case of Europe's new weed beets, highlighted in chapters 1 and 6, is perhaps the most dramatic of the lot. These plants have resulted in millions of dollars of losses to farmers and other members of Europe's sugar industry. Some other weedy hybrid lineages have had equally dramatic consequences. For example, California's weedy rye, a derivative of natural hybridization between *Secale cereale* and *S. montanum,* has become such a problem that California "farmers have abandoned efforts to grow rye for human consumption" (National Academy of Sciences 1989).

Increased aggressiveness in hybrid lineages has also apparently resulted in the evolution of new invasives. One example is the case of *Rhododendron ponticum,* a particularly invasive species of the British Isles (Gray 1986), which has had such an impact that it is one of the key species involved in the conversion of native heathlands into exotic shrub-forest complexes (Mitchell et al. 2000). As pointed out above, this species has enjoyed an expanded niche in Britain and Ireland, moving into much colder areas than it occupies in its native range, because of introgression of cold-tolerant alleles from the North American ornamental *R. catawbiense* (Milne and Abbott 2000). Interestingly, in its native range, where such introgression has not occurred, *R. ponticum* is not particularly invasive.

In addition to the evolution of new weeds and new invasives, gene flow from cultivated species has also resulted in the kind of evolution that makes nasty plants even more difficult to manage. For example, johnsongrass, *Sorghum halepense,* is one of the world's worst weeds (Holm et al. 1977) but is particularly aggressive in North America. Features of the weed's morphology and development there have led weed evolutionists to hypothesize that introgression from grain sorghum, *S. bicolor,* accounts for the evolution of that increased weediness (e.g., de Wet and Harlan 1975; P. Morrell, personal communication) "Crop mimicry" (discussed in chap. 5) is another example of such evolution; human hand-weeding selects for weeds that are increasing visually similar to the crop (Barrett 1983). If the weed is more similar, it is increasingly difficult to recognize. For example, wild rice taxa are serious weeds of rice and other crops (Holm et al. 1997). For hundreds of years they have been a problem because their seedlings resemble cultivated rice. To overcome the problem of distinguishing between wild and cultivated seedlings of rice, Indian plant breeders developed purple-leafed rice cultivars whose seedlings could be easily identified when growing among the green-leafed weeds. Shortly after their introduction, natural hybridization between crop and weed, followed by selection by hand-weeding favoring purple-leafed morphs, resulted in the establishment of purple-leafed weeds that were almost identical to the crop (Dave 1943, Oka and Chang 1959, Harlan et al. 1973, Parker and Dean 1976). Introgression from pearl millet into populations of its close relatives in Africa represents another example of gene-flow-mediated crop mimicry (Brunken et al. 1977, Barrett 1983).

It is clear that hybridization between wild plants and their domesticated relatives has resulted in some plants that have created hardship for humankind and others that have disrupted natural ecosystems. But is there any reason to believe that spontaneous hybridization involving cultivated species creates problem plants more frequently than hybridization in general? Surprisingly, it appears that the answer is yes. The few thousand domesticated plant species make up only about 1 percent of the approximately 250,000 known higher plants (Wilson 1992). But when Ellstrand and Schierenbeck (2000) created an inventory of well-studied examples of the evolution of plant invasiveness following hybridization, 25 percent of their cases listed (7 of 28) have a cultivated taxon as one of the hybridizing parents, and that list does not include all of the examples presented in tables 9.2 and 9.3. The chances of a domesticate-wild hybrid derivative becoming a problem plant are an order of magnitude greater than if both hybridizing parents are wild species.

Increased Risk of Extinction?

The other potential concern for managers of wild plants created by hybridization between domesticated plants and their wild relatives is the increased risk of extinction of wild taxa. For example, *Ulmus pumila,* the Siberian elm, is native to eastern Asia, where it has been used for hundreds of years for timber, shelter, and fodder. It was introduced as an ornamental into Spain in the 1500s (Cogolludo-Agustín et al. 2000). Since that time apparent hybrids between that species and the native field elm, *U. minor,* have become increasingly common. Isozyme analysis has confirmed that plants of intermediate morphology were descendants of natural hybrids between the two species (Cogolludo-Agustín et al. 2000). Extensive natural hybridization with the ornamental together with losses due to Dutch elm disease have made genetically pure individuals of the native species increasingly rare. Indeed, the ornamental has considerable resistance to Dutch elm disease, and it is not surprising that the current plants of hybrid ancestry "appear to be nearer to *U. pumila* than *U. minor*" (Cogolludo-Agustín et al. 2000).

The overall impression of experts in the field is that the hybridization-mediated extinction of the progenitors of domesticated plants is not rare (e.g., Small 1984). Extinction by hybridization with domesticated species has been implicated in the extinction or increased risk of extinction of the wild relatives of several cultivated plants, including capsicum pepper, date palm, hemp, maize, and sweet pea (reviewed by Small 1984). But that conclusion is often based on the observations that those progenitors are now missing and that weedy forms of the crop are hybrid lineages that appear to carry a shadow of the genetic heritage of the wild species. However, the topic has enjoyed more speculation than data collection.

Table 9.4 lists ten cases having more than anecdotal evidence that gene flow from a cultigen is endangering one or more wild populations. One of the possible reasons that data are hard to come by is that the process of extinction under hybridization may be very rapid (e.g., Wolf et al. 2001), perhaps taking only a handful of generations, which may be much less than a century (or even a decade) for annual or biennial species. Except for the well-documented study of the demise of Taiwanese wild rice discussed above (Kiang et al. 1979, Oka 1992), these studies typically provide some data that confirm that hybridization has occurred and document the extent of that hybridization.

The topic is ripe for more research. What are the relative fitnesses of the hybrids and hybrid derivatives compared with the genetically pure plants? What fraction of the progeny produced by the genetically pure plants are currently sired by the cultivated species or by a hybrid derivative? How does the frequency of hybrid derivatives change over time? Do those changes correlate with changes in the community of organisms that eat these plants? Do historical collections exist (e.g., herbarium materials) that permit the analysis of the rate of introgression over time?

It is important that these questions be answered. Indeed, hybridization between crops and their wild relatives has probably resulted in the extinction of many species since the Agricultural Revolution thousands of years ago (Small 1984). The extraordinary speed of the extinction process discussed elsewhere in this book may be one reason those extinctions have been overlooked. But attention has recently begun to focus on extinction by hybridization as an important conservation problem (e.g., Allendorf et al. 2001, Levin et al. 1996, Rhymer and Simberloff 1996).

In fact, the new Agricultural Revolution of genetic engineering has spotlighted gene flow and hybridization as potential opportunities for the escape of engineered genes, which might create more problems of the sort discussed in this chapter. Will engineered plants be more or less of a gene-flow problem? That question is considered in the next two chapters. It is considered in chapter 10 for "The Case of the Bolting Beets," and in chapter 11 more broadly for the whole class of engineered plants that have been released into the field as well as for those that are coming soon.

The Case of the Bolting Beets

Part III. The Cloudy Crystal Ball

The future ain't what it used to be.

—ATTRIBUTED TO YOGI BERRA, BASEBALL PLAYER

The mystery was solved, but the case wasn't closed.

The genetic detectives had collected evidence that would compel any jury to conclude that northern Europe's weed beets are hybrids and the descendants of hybrids between wild beets and sugar beets in southern Europe's seed production fields. Furthermore, it was clear that hybridization had been constantly adding new individuals to Europe's populations of weed beets. Therefore, protocols were developed to minimize contamination in seed production fields and to improve weed beet control (Longden 1993). In fact, beet seed production quality standards now require rigorous testing for bolters, so that if more than 0.1 percent bolting is detected in a seed lot, the entire lot is discarded (Chazallon 2000). Nonetheless, problems with weed beets continue, and the weeds are even spreading into Eastern Europe (D. Bartsch, personal communication).

But the detectives raised a new worry: Would plans for improving sugar beets also end up improving the weed? That is, they worried that genetically engineered sugar beets might pass on their new genes to the weeds—and they worried about the consequences of those transgenes in wild populations.

The answer to the first part of that concern was clear. If sugar beets, transgenic or not, naturally mate with wild beets, crop genes, engineered or not, would inevitably find their way into weed populations. We can imagine at least three pathways. First, unless steps were specifically taken to prevent wild beets from mating with transgenic plants grown for seed, hybridization would occur as it has for years, as described in chapter 1, with transgenic weed beets evolving as contaminants directly during seed production (albeit at a low rate, with the new rigorous testing described above). Second, a tiny percentage of the genetically pure sugar beet crop bolts spontaneously. Thus, we would expect to find rare transgenic purebred bolters in northern Europe's sugar beet fields—amid many willing potential mates in the now self-sustaining weed beet populations. Third, not all beets that are planted are harvested, and those that aren't harvested may flower. Seeds can fall outside the cultivated field. Harvesting machinery might fail to collect every plant. A farmer might even decide against harvesting his or her fields if the price of sugar falls to the point that harvest is unprofitable. A surprising fraction of beets left in the soil can survive through the winter; under the right conditions that fraction can be almost as high as 90 percent (Pohl-Orf et al. 1999). And winter's cold is the trigger that causes them to bolt. Predicting this part of the future is easy. Unless specific steps are taken to prevent mating, genes in Europe's sugar beets, engineered or not, should move with ease into populations of weed beets. In fact, from the data reviewed in chapter 6, it appears that cultivated beet genes will flow into any nearby related wild beet populations, whether they are the weed beets of northern France, the wild beets of Italy's Adriatic coast, or the weedy *Beta macrocarpa* of California's Imperial Valley.

But the answer to the second part of that concern, the impact of transgenes in wild populations, is much less clear. The significance of any new allele in the wild depends on how that allele is expressed in the wild. For a noxious weed, alleles of greatest economic concern are those that would make that weed more difficult to control. Just look at the effect of the bolting allele in beets. Short of conducting experiments on transgenic hybrids, the best prediction that can be made about consequences of transgenes in natural populations depends on the knowledge of the expression of the transgene, the biology of the plant expressing the transgene, and the environment in which those plants grow (Ellstrand and Hoffman 1990). Clearly, different transgenes will have very different impacts based on the traits they confer.

So what traits are genetic engineers adding to sugar beets? Like other important crop plants, sugar beet has been the object of considerable genetic engineering in many laboratories, public and private. Although it is not possible to ascertain all of the types of transgenic sugar beets created in the world's laboratories, it is possible to obtain public records of applications for the intentional release of transgenic plants into the environment. Lists of approved field tests are particularly useful for our purposes because they represent genotypes that have already proven themselves in the laboratory and greenhouse and may be on their way to being grown commercially. Also, a small number of transgenic plants are now beyond the field-testing stage; they have been deregulated and can be grown commercially.

The European Commission's "GMOs in Food and Environment" website has a database of deliberate field trials of transgenic organisms in sixteen European countries (http://biotech.jrc.it/deliberate/gmo.asp). According to that database, as of 10 January 2000, 273 field trials had been conducted on genetically engineered table beets, leaf beets (Swiss chard), sugar beets, and fodder beets in twelve of those countries. (Because of the European controversy over transgenic plants, very few field trials have been conducted in those countries since that date.) Of those countries, the one with the greatest number of beet trials was no surprise: France, with sixty-seven.

Weeds are such an important problem for beets that herbicide resistance has been the product of choice for the genetic engineers of sugar beets. Of the field trials listed in the European Commission (EC) database, more than one-third involved beets specifically engineered for resistance to the herbicide glyphosate (also known as Roundup®), and more than one-quarter involved beets engineered specifically for resistance to another herbicidal compound, phosphinothricin (also known as glufosinate or BASTA®). Many of the other field trials involved beets engineered for multiple traits ("stacked genes"), usually including resistance to at least one herbicide. For example, several field trials involved beets resistant to both glyphosate and glufosinate. Beyond herbicide resistance, the second most common engineered trait among the EC field trials was resistance to the virus that causes the important sugar beet disease rhizomania, caused by the beet necrotic yellow-vein virus (BNYVV). That trait was usually stacked with genes conferring other traits. Although beets with stacked genes made up a minority of the examples on this list, "stacks" of three or more traits were not unusual.

Another source of information on deliberately released transgenic crops is a database maintained by Information Systems for Biotechnology (ISB) of Virginia Tech University. That database of all applications for field trials and all petitions for deregulation of transgenic plants in the United States is part of the larger ISB website (http://www.nbiap.vt.edu). At the end of July 2001, the ISB database listed 123 applications for field-testing transgenic beets in the United States. The agency receiving the applications, the Animal and Plant Health Inspection Service of the United States Department of Agriculture (USDA-APHIS), is responsible for oversight of all field releases of transgenic plants in the United States. Because one field test application often lists multiple release locations, the total field tests conducted on transgenic beets in the United States number in the hundreds. As in Europe, herbicide-resistant lines have been the primary product of beet genetic engineers in the United States. Of the beet applications filed with USDA-APHIS, more than half involved glyphosate resistance and more than one-third involved glufosinate resistance. Only nine applications (less than 10 percent) were transgenic for another trait, virus resistance (mostly resistance to BNYVV). In contrast to the European list, not a single application for field-testing beets involved stacked traits.

Although plenty of field tests have been conducted on transgenic sugar beets, both in the United States and elsewhere, only two types have yet been deregulated for commercial production, according to the online "agbios" database of Agriculture and Biotechnology Strategies (Canada), Inc. (http:// agbios.com/dbase.php?action=ShowForm). One line, T120-7, a product of AgrEvo (now known as Aventis Crop Science), is glufosinate resistant. That line has been deregulated in the United States (Petition #97-336-01p). Deregulation in the United States means that a transgenic plant is treated like any other crop; it can be grown for any purpose without federal regulatory oversight. The same line has also been approved for human food in Japan and Canada and for animal feed in Japan. The other sugar beet line, GTSB77, created by Novartis and Monsanto, is resistant to glyphosate. That line has also been deregulated in the United States (Petition #98-173-01p).

Did regulators consider hybridization to be a problem in their decisions regarding transgenic beets? The decision documents of USDA-APHIS are available from the ISB website (http://www.nbiap.vt.edu) and offer a glimpse into the American decision-making process. Let's consider two case studies from the United States: the most recent field test permit for transgenic sugar beet and the most recent petition for deregulated sugar beet.

Since 1998 most American field tests have been conducted under the USDA-APHIS notification system. That system is an expeditious one, involving minimal paperwork (essentially a brief form), no Environmental Assessment, and a rapid decision by USDA-APHIS (APHIS 1997b). USDA-APHIS considers the application and, if it appears to be in order, acknowledges the notification without creating a document to evaluate the application.

Therefore, it has been years since a decision document has been prepared for a sugar beet field test in the United States. Betaseed, Inc., submitted the most recent sugar beet application for a field test permit in 1997. After conducting an Environmental Assessment, USDA-APHIS issued a permit (Permit #97-044-02R) a little less than three months after receiving Betaseed's application.

Betaseed's transgenic beet was engineered with a gene conferring resistance to a virus that causes disease in sugar beets. The decision document (APHIS 1997a) does not report the specific disease or virus, because the applicant claimed both details to be "confidential business information." Virus resistance has obvious agronomic value. The trait is also one that might benefit wild plants in locations where the disease is present. Under USDA-APHIS regulations, plants in field tests must be managed to prevent mating with cross-compatible plants. The USDA-APHIS decision document of Betaseed's application raises the issue of the escape of the transgene into natural populations:

> To prevent dissemination of pollen during the field trial, sugar beet plants that bolt will be removed from the field test site and destroyed or stored in an approved containment facility. Should any seed accidentally be produced on the plants at the site, any volunteer seedlings will be detected and destroyed prior to flowering the following growing season. The natural distribution of wild species of *Beta* . . . is largely along the Atlantic coasts of Europe and North Africa and in the Mediterranean. . . . Therefore, cross-pollination of wild populations of *Beta* by the transgenic sugar beets in this experiment is not possible. . . .
>
> Occurrence of wild beets has been documented in California. . . . In that case, the sugar beets had evolved into annual plants and had become weeds. There is no reason to believe that the genetic modification in the transgenic sugar beet

line would cause these sugar beets to become annuals or otherwise become weeds. No *Beta* species are listed on the States' list of noxious weeds, and other than the case of the wild annual beets in California, *Beta vulgaris* is not listed as a serious, principle [*sic*], or common weed in the U.S.

If executed thoroughly, Betaseed's plans to destroy bolters and volunteers before they set seed would very likely be sufficient to prevent the escape of transgenes from the Idaho test site. Interestingly, this 1997 analysis neglected to mention the decades-old weed beet problem of Europe and the discoveries of the early 1990s that weed beet in Europe was the result of hybridization between the crop and wild populations. It was not until two years later that California's wild beets were shown to be not only feral cultivars but also *B. vulgaris* ssp. *maritima* and *B. macrocarpa* (Bartsch and Ellstrand 1999).

A more thorough analysis was conducted a few months later by USDA-APHIS on a petition from Novartis Seeds and the Monsanto Company requesting deregulation of transgenic glyphosate-resistant sugar beet. Glyphosate is the ingredient in Monsanto's herbicide Roundup®, an herbicide capable of killing all plants (except two recently evolved resistant races; Prather et al. 2000) but one that presents little, if any, toxicity to most vertebrates, including humans (Franz et al. 1997). An "environmentally friendly" herbicide, it "is rapidly biodegraded into natural products by microorganisms present in both soil and water" (Franz et al. 1997). Glyphosate is widely used as an herbicide for a great variety of crops as an alternative to tilling, thus reducing soil erosion. A relatively safe herbicide that kills almost all plants has only one drawback: you can't use it to control weeds after your crop emerges unless you can be sure that it doesn't come into contact with your crop. A glyphosate-resistant crop would provide the benefit of complete weed control both before and after the crop emerges. Obviously, a weed that received that trait by hybridization would also enjoy a big benefit if exposed to the herbicide.

The USDA-APHIS determination document (APHIS 1998) considered hybridization and the consequences of glyphosate-resistant weeds in some detail. The treatment of those topics is illuminating (my comments appear in brackets):

> Movement of the transgenes via pollen from line GTSB77 to other members of the Beta section is species-specific. Movement of the transgenes to *B. vulgaris* subsp. *atriplicofolia,* subsp. *orientalis,* subsp. *maritima,* and subsp. *patula* is not likely since these plants are not found in the Americas. Movement of the

transgenes from GTSB77 to potentially sexual compatible species is likely in two localities. Sugar beet plants escaped from past commercial cultivation in the San Francisco Bay area and persist to this day. However, sugar beets are no longer in commercial production in the Bay area, and thus transgene movement via pollen to these plants is highly unlikely.

The situation in the Imperial Valley of California is more complicated. There are free living sugar beets that have escaped cultivation and have persisted (McFarlane, 1975; Johnson and Burtch, 1958). These plants are a minor weed problem in this area. Movement of the transgenes from GTSB77 to these plants is likely. The other plant in the Imperial Valley that commercial sugar beets could potentially successfully pollinate is subsp. *macrocarpa* [now known as the separate species *Beta macrocarpa*]. There appears to be conflicting evidence on whether commercial sugar beets can pollinate subsp. *macrocarpa*. Bartsch et al. (1996) [not listed in the document's references] has suggested based on isozyme analysis that introgression of genes from commercial sugar beets has occurred. Lee Panella of the Sugar Beet Crop Germplasm Committee provided two lines of evidence that gene flow between these two plant populations is not likely. First, commercial sugar beets and subsp. *macrocarpa* flower at different times. Second, in greenhouse crosses between these plants, most F1 hybrids are sterile and the F2 hybrids had very disturbed growth patterns and genetic ratios. In greenhouse crosses, successful hybrids occurred only when sugar beets were the female parent.

Some scientists (Boudry et al., 1993; Bartsch and Pohl-Orf, 1996) have questioned whether movement of herbicide tolerance genes from commercial sugar beets to sexually compatible relatives poses an environmental risk. APHIS believes that if and when the glyphosate tolerance trait moves from GTSB77 to other sexually compatible *Beta* sp. this will not have a significant impact. If glyphosate tolerant individuals did arise through interspecific or intergeneric hybridization, the tolerance would not confer any competitive advantage to these plants unless challenged by glyphosate. This would only occur in managed ecosystems where glyphosate is applied for broad spectrum weed control, or in plant varieties developed to exhibit glyphosate tolerance and in which glyphosate is used to control weeds. As with glyphosate tolerant sugar beet volunteers, these individuals, should they arise, would be controlled using other available chemical means. Hybrids, if they developed, could potentially result in the loss of glyphosate as a tool to control these species. However, this can be avoided by the use of sound crop management practices by not using the same herbicide every year.

The petition was approved six months after it was received. In contrast to the Betaseed permit application analysis, the petition analysis did include mention (if only brief) of crop-wild hybrids as a weed problem in European production fields. But it did not mention that wild beet is an important weed of sugar beets and certain other crops in the Imperial Valley (Robbins et al. 1970). Given that fact plus the recent finding that about two out of every hundred *B. macrocarpa* plants in the Imperial Valley have introgressed sugar beet alleles (Bartsch and Ellstrand 1999; see chap. 6), the USDA-APHIS analysis now sounds a bit optimistic about the likelihood and significance of crop transgene introgression in that particular location.

Still, whether the evolution of glyphosate-resistant beets poses a "significant impact" may be in the eye of the beholder. Predicting the impact of glyphosate-resistant weed beets in California's Imperial Valley involves more than just considering the impact on sugar beet growers. Consider the following interchange I had with an Imperial Valley sugar beet grower in 1999, soon after herbicide-resistant sugar beets were deregulated.

The grower called with a complaint. After years of promises that genetic engineering would deliver herbicide-resistant sugar beets, the companies producing them were reluctant to sell them for use in the Imperial Valley. Someone had suggested to the grower that he call me for clarification. I explained that, under hybridization and intense selection by the herbicide, it wouldn't be long before herbicide-resistant weed beets would evolve and present a new problem.

He already understood the situation. "But the other growers and I are willing to enjoy a few good years of weed-free production until those weeds evolve. Then I'll switch to another herbicide."

"But," I countered, "don't you have neighboring farmers in Imperial who grow vegetable crops other than sugar beets and use that herbicide, and have good reasons for using it: to control weed beets and other weeds in their fields?"

"Sure."

"Then are you willing to force your neighbors to give up that herbicide because of the new weed beets you've created?"

His response was quick: "I get it. Thanks."

Thinking about impacts involves thinking outside of the box. The regulators, the companies, and the growers all initially considered the primary transgenic trait (herbicide resistance), the organism (sugar beet), and the immediate environment in which the transgenic sugar beets were to be deployed (sugar beet production fields). And because of the attention that had been given to the gene-flow issue, they had considered the possibility of hybridization. But weed seeds aren't confined within the bounds of a sugar beet production field. Weed beet seeds can float down roadside ditches swollen with irrigation runoff, move considerable distances in the mud sticking to tractor tires, or enjoy a lift from southern California's intense Santa Ana winds. But it wasn't until commercialization that the companies producing the transgenic crop came to realize that in certain locations, hybridization with weeds would result in the evolution of an herbicide-resistant weed and the obsolescence of the very herbicide they were seeking to promote. Since California is the only location in the United States with wild beet populations, it was practical to prevent the sale of herbicide-resistant beets in that state.

The foregoing discussion of the environmental impacts of transgenic sugar beet hybrids is largely hypothetical. But there's nothing like the real thing. While scientists in the biotechnology industry, the regulatory community, and public interest groups have speculated—and sometimes argued—about the potential environmental impacts of natural hybrids between engineered beets and their wild relatives, other scientists set out to measure the actual performance of the hybrids. The first such studies asked the obvious question: Do crop-wild hybrids inherit and express transgenes as they would any other trait?

Bartsch and Pohl-Orf (1996) crossed transgenic sugar beets with nontransgenic beets. The four sugar beet lines used as male parents for the crosses were stacked transgenics, homozygous for resistance to the viral disease rhizomania (*cp* gene), for resistance to the herbicide glufosinate (*bar* gene), and for a marker gene for antibiotic (kanamycin) resistance (*nptII* gene). The nontransgenic beets used as the female parents were accessions from wild populations of sea beet (*B. vulgaris* ssp. *maritima*) as well as one red beet cultivar and one spinach beet (Swiss chard) cultivar. The hybrid offspring were tested for glufosinate resistance by placing a drop of the herbicide solution on one of the cotyledons to determine whether the stacked engineered construct had been transmitted to the hybrid offspring. The response was compared with the cotyledon damage created on nontransgenic controls by the same test. As expected, all of the hybrid seedlings "proved to be resistant against the herbicide." The transgene was stably inherited, and the wild genome did not appear to affect the expression of the herbicide-resistance transgene.

Subsequently, Dietz-Pfeilstetter and Kirchner (1998) used a similar experimental system to expand on those findings. The sugar beet lines used for the parents of the hybrids had the same set of stacked transgenes as those lines used by the previous study. The wild beet parents were two different accessions of sea beet, *B. vulgaris* ssp. *maritima*. PCR analysis revealed that the hybrids all received both the *cp* gene and the *bar* gene as expected by Mendelian transmission and transmitted them to their F_2 progeny in the expected 3:1 ratio. The hybrids were herbicide resistant; spray application of glufosinate at the recommended agricultural rate had no detrimental effect. To test for rhizomania resistance, plants were inoculated with the virus, leading to a complex result. One of the wild beet accessions already had a high degree of resistance to the disease, and its hybrids were no more resistant than the wild parent; the other accession was susceptible to the disease, and its hybrids enjoyed significantly higher virus resistance. The plants in the aforementioned experiments were grown in the greenhouse.

But how do the hybrids behave under field conditions? Interestingly, field studies involving transgenic beet hybrids are extremely rare, possibly because of fears of transgene escape from bolting experimental beets. Madsen et al. (1998) compared the competitive abilities of hybrids between a sea beet accession and transgenic glyphosate-resistant sugar beet with the pure lines. Seeds of the different types were sown in a Danish field in different densities both in monoculture and in paired mixtures. Based on aboveground biomass, "the hybrid line showed the highest competitive ability" of the three types, which the authors of the study attributed to hybrid vigor (Madsen et al. 1998).

Is it hybridity or is it the transgenic genotype that explains the fitness boost of transgenic hybrids? A field study (Bartsch et al. 2001) was conducted to address that question, comparing the fitness of nontransgenic hybrids with that of transgenic hybrids that were otherwise as genetically similar as possible. The authors created nontransgenic and transgenic hybrids by crossing transgenic sugar beet or its nontransgenic parent with a close cultivated relative, Swiss chard, which served as a model for wild beet. The model is a reasonable one because "Swiss chard is weedy and apparently naturalizes easily" (Bartsch et al. 2001). For example, morphological and allozyme analysis of California wild beets provided substantial evidence that many populations evolved from naturalized Swiss chard (Bartsch and Ellstrand 1999). The sugar beet line used in this experiment was transgenic for the same stacked genes (rhizomania resistance, glufosinate resistance, and kanamycin resistance) as the Bartsch and Pohl-Orf (1996) and Dietz-Pfeilstetter and Kirchner (1998) studies detailed above. Three types of plants—transgenic hybrids, nontransgenic hybrids, and the parental Swiss chard cultivar—were grown at a German field site. The plants were grown under either low or high levels of exposure to the virus that causes rhizomania with one of three levels of competition from the common weed *Chenopodium album*. Relative fitness of the three types, estimated from plant biomass, did not vary with weed competition but showed different patterns according to virus exposure levels. Nontransgenic hybrids always had the highest fitness of the three for the low virus exposure treatments; the transgenic hybrids, for the high exposure treatment. Unexpectedly, the transgenic hybrids had a very low bolting rate relative to both their nontransgenic counterpart and their Swiss chard parent.

These experiments have shown that a genetically engineered sugar beet will pass on its transgenes like any other gene to its hybrid children, and their grandchildren appear to inherit those genes in the manner that Mendel predicted. For the engineered genes studied, transgenic expression appears to be unaltered by coexistence with alleles having a wild ancestry. And those engineered genes confer the expected advantage to the hybrid in response to the appropriate environmental stress (virus resistance in the presence of high levels of virus, herbicide resistance in the presence of herbicide). These advantages, coupled with the hybrid vigor observed in these experiments, suggests that transgenes will persist and spread through weed beet populations.

Will they make weed beets weedier? Certainly, if herbicide resistance spreads, farmers will be compelled to switch to a different herbicide. At best that change will be a headache for the farmers already beleaguered by weed beets. But the situation for other transgenes is less clear. Even the case of virus resistance is hard to predict. Virus resistance confers an advantage only at high infection levels. How will that advantage play out? While it is easy to predict that the gene for resistance should sweep through populations suffering high levels of the disease, it is not easy to predict whether that advantage will make weed beets more aggressive or difficult to control. A plant's niche is determined by the complex interplay of its genotype and its environment.

And sometimes that genotype is expressed in unexpected ways. The unpredicted change in bolting pattern observed in transgenic hybrid beets by Bartsch et al. (2001) represents both good news and bad news. The good news is that because transgenic hybrids bolt less frequently, they are less likely to pass on their genes than their more frequently bolting nontransgenic parents. This difference will tend to slow the spread of transgenes.

The bad news is that these data appear to demonstrate that transgenes can have unintended and unpredictable effects on a plant's phenotype. Such pleiotropic effects are well known though poorly understood for "conventional" genes but have been rarely considered for transgenes. Given that transgenes in crop-wild hybrids have different alleles to interact with and are expressed in an environment different from that of the crop field trials, pleiotropic effects of transgenes in wild populations might go unnoticed until those transgenes become established in wild populations. Pleiotropic effects are not necessarily good, nor are they necessarily bad; they are just unpredictable. The only thing that we can predict with certainty about the future of the bolting beets is that the future is going to be somewhat different from the past.

Interestingly, although transgenic sugar beets were deregulated in 1998, they are not currently grown for commercial production anywhere in the United States or elsewhere in the world. The reason has nothing to do with environmental risks. Because of general public concerns, particularly health-related ones, about genetically engineered crops, major sugar users in the food industry—companies like Hershey Food Corporation and M&M/Mars—are unwilling to buy sugar from transgenic beets until prevailing attitudes change (Kilman 2001). The existence of an informal boycott based on the impact of genetically engineered sugar on human health is ironic because sugar is the extremely pure biochemical sucrose, and while there may be valid reasons for questioning genetically engineered crops and products, sucrose is sucrose, whether produced by sugar beet or by sugarcane, transgenic or not.

What will be the final chapter in "The Case of the Bolting Beets"? Haze, heat, desiccation, and white, sepulchral, alkaline soil—California's Imperial Valley is about as different from northern France as you can get. The valley is one of the hottest places in North America, so hot most of the year that sugar beets are planted in winter. It is also one of the driest places in North America, so dry that the beets depend on irrigation. The valley's soil is so alkaline that the primary weeds are those species closely related to beets, including *B. macrocarpa*.

Given that the United States is a champion of crop biotechnology and that the Imperial Valley remains a center of sugar beet production, the valley may well be the setting for the last chapter in "The Case of the Bolting Beets." Sacks of transgenic herbicide-resistant sugar beet seeds wait to be planted, and many California farmers await consumer acceptance of genetic engineering so that they can try out these plants. While the companies that created these seeds might still be reluctant to sell them to California growers, other companies that are developing herbicide-resistant beets might not be so prudent. And *B. macrocarpa* weeds are waiting to romance these exotic mates. How many generations of hybridization and selection would have to occur before herbicide-resistant *B. macrocarpa* became a problem to Imperial Valley vegetable growers?

We're not talking "killer tomatoes" here. The worst possible scenario for a new weed with resistance to a major broad-spectrum herbicide would be on the order of no more than hundreds of millions of dollars; that figure would include depressed yields of crops suffering competition with the new weed, increased costs for controlling the new weed, and lost revenue to the company producing that herbicide once it becomes ineffective.

At the other extreme is the possibility that the transgene will never successfully introgress into natural populations. As pointed out elsewhere in this book, even with repeated hybridization, certain alleles have a difficult time introgressing from one population to another. Or it is possible that transgenic beets, when they are released in the Imperial Valley, will be genetically engineered with special constructs to frustrate introgression (e.g., Gressel 1999; see chap. 12). Finally, the companies selling transgenic sugar beets may continue to consider the possibility of introgression to be too much of a risk and continue their ban on selling transgenic beets in regions where wild relatives live.

The most likely scenario falls between those extremes, especially if resistance management is practiced. First, the evolution of resistance can be delayed or prevented. Herbicide rotation is one strategy for frustrating the evolution of herbicide resistance (Gressel 1991). Another well-known crop management practice is crop rotation (Rubin 1991, Prather et al. 2000). The conditions under which alternative crops are grown may include differences in planting time, differences in the competitiveness of the alternative crop, and different cultivation conditions, in addition to the use of other herbicides. These differences may act to decrease frequencies of preexisting weed species.

Should herbicide-resistant weed beets evolve, the primary resistance management methods should be to rotate herbicides or to employ specific strategies for preventing herbicide-resistant weeds from spreading (Prather et al. 2000). Rotating with alternative herbicides may be more expensive, less environmentally benign or less effective in controlling a wide variety of weeds, and, according to at least one expert, "bound to fail in the long run" (Rubin 1991). Herbicide rotation also has the downside of not accruing the benefit afforded by the transgenic crop. Preventing the establishment and spread of herbicide-resistant weeds requires the early identification of such plants by monitoring and their subsequent eradication or vigilant hand-weeding. With continued strong selection pressure from the herbicide, it is unlikely that resistant weeds constantly appearing from new bouts of natural hybridization will be entirely and permanently controlled. Once resistance becomes established in seed banks and multiple sites, the second line of resistance management, crop rotation, will be necessary. The evolution of herbicide-resistant weeds is one more agronomic challenge to farmers. It is certainly not clear how big that challenge will be.

An experiment is waiting to happen.

Or is it? Sugar beets are falling into disfavor in the United States and elsewhere. Sugar beet–processing plants are shutting down. Globally, sugar production is shifting to sugarcane, and other sweeteners are replacing sugar. Maybe "The Case of the Bolting Beets" is already over and the last chapter has already been written. Perhaps the slow demise of beets as a source of sugar will mean that a romance between another cultivated-wild pair will provide the definitive case study as to whether engineered genes will pose a special problem for wild plants. That is the topic of the next chapter.

CHAPTER 11

The "Special" Case of Genetically Engineered Plants?

> With one or two exceptions, transgenic crop plants are unlikely to become weeds or to transmit new genes to related nearby weeds.
>
> —PETER R. DAY, 1987

> Existing ecological theory and emerging research data suggest that the massive planting of transgenic monocultures can create critical environmental impacts ranging from gene flow between transgenic crops and wild relatives, the creation of super-weeds and the rapid development of insect resistance, to impacts on soil fauna and nontarget organisms.
>
> —MIGUEL A. ALTIERI, 2000

There's been lots of talk about the health, social, economic, and environmental impacts of genetically engineered plants. Of the environmental concerns, those associated with the escape of engineered genes (also known as "transgenes") into the populations of wild relatives have received the most attention. Indeed, almost every general treatment of the environmental impacts of plant biotechnology gives some consideration to gene flow (e.g., Carpenter 2001, Colwell et al. 1985, Cook 2000, Dale et al. 2002, Ervin et al. 2000, Goodman and Newell 1985, Hails 2000, Keeler and Turner 1990, Marvier 2001, McHughen 2000, National Academy of Sciences 1989, 2000, 2002, Nickson and Head 1999, Pretty 2001, Rissler and Mellon 1996, Scientists' Working Group on Biosafety 1998, Snow and Moran-Palma 1997, Tiedje et al. 1989, Traynor and Westwood 1999, Van Aken 1999, Wolfenbarger and Phifer 2000). Interestingly, problems associated with hybridization between conventional crops and their wild relatives received scant attention until potential gene-flow problems were described for transgenic crops (but see Barrett 1983, Small 1984).

Four primary concerns have been raised about transgenes in natural populations. The issue most frequently raised is whether transgenes will confer a benefit to weedy relatives, resulting in the evolution of "superweeds" that are very difficult to control or new invasive lineages that overrun and disrupt natural ecosystems. Second is the question of whether the wild relatives of transgenic crops will suffer an increased risk of extinction by hybridization with those crops. Third is the fear that gene flow from transgenic crops poses a threat to the genetic diversity stored in wild populations. Those three concerns correspond closely to topics already covered in chapter 9 for conventional crops. Finally, there is the assertion that gene flow from transgenic crops to natural populations is a hazard in itself as a kind of genetic pollution.

This chapter explores the question of whether gene flow from transgenic crops poses special environmental problems relative to gene flow from the products of traditional plant improvement. That is, will replacing a traditionally bred variety with one that is transgenic significantly increase (or decrease) problems created by gene flow? Before examining that question, it is first important to consider briefly what aspects of transgenic plants set them apart from conventional crops. Specifically, I consider those aspects that might influence the impact of transgenic-wild hybridization.

What Is It about Transgenes That Might Influence the Impact of Crop-Wild Hybridization?

Are genetically engineered plants "special" with regard to the process by which they are created? Engineered genes are those that are intentionally inserted into the hereditary material of an organism by the methods of biotechnology rather than by the traditional breeder's method of sex. The source of the inserted genes can range from very close relatives to the most distant ones. Indeed, at one extreme the source can be the organism itself; its genes can be extracted and reintroduced into its genome in the form of one or more extra copies (e.g., Alvarez et al. 2000). At the other extreme are alleles introduced into plants from organisms as diverse and distant as jellyfish, bacteria, viruses, and humans. Engineered genes can also be artificial DNA sequences created in the laboratory, unavailable from any organism. In itself the methodology of genetic engineering is "special" only in that it has the potential to generate genetic combinations that are difficult or impossible to create by traditional breeding techniques. With regard to superficial phenotype, certain products of genetic engineering could have been created by traditional breeding. For example, although virus resistance is a common engineered trait, many virus-resistant crop varieties have been created without engineering (Grumet 1995, Hadidi et al. 1998). Still, the mechanism of virus resistance differs between the two methods. Virus resistance introduced into a cultivar by traditional breeding comes from preexisting variation in the crop or relatives of the crop. But in transgenic plants the most frequent source of virus resistance is a gene taken from the offending virus itself (Powell-Abel et al. 1986). However, there is a growing list of transgenic phenotypes that would be impossible or nearly impossible for traditional breeding to create. A spectacular example is plants that glow green under ultraviolet light, a trait based on a protein created by a jellyfish gene (Stewart 2001).

Does the transmission genetics of transgenes in hybrids differ from crop genes? For transgenic crops to be worthy of commercial production, the transgene must be transmitted and expressed in a stable fashion. Thus, after they generate new genotypes expressing the desired trait, the creators of engineered plants, just like traditional breeders, select plants that transmit and express the traits reliably. Because wild relatives are so closely related to the crops with which they hybridize, we would expect that stability to be maintained in their hybrid progeny. The experiments described in the last chapter (Bartsch and Pohl-Orf 1996, Dietz-Pfeilstetter and Kirchner 1998) demonstrated that sugar beet transgenes were transmitted to and expressed in first- and second-generation crop-wild hybrids. Similar experimental work has shown that crop transgenes are transmitted stably and expressed predictably in the progeny of other crosses between crops and their wild relatives. For example, transgenic glufosinate resistance was transmitted to and expressed in F_2s derived from crosses between cultivated rice and the weed red rice (Oard et al. 2000). Likewise, backcrosses as advanced as B_3s derived from hybrids of transgenic *Brassica napus* crossed with wild *B. rapa* (= *B. campestris*) expressed glufosinate resistance inherited from the transgenic ancestor (Snow et al. 1999).

Transgene transmission sometimes varies from Mendelian expectations in hybrids derived from crosses between a transgenic parent and a wild taxon. Oard et al. (2000) measured transgenic glufosinate-resistance segregation patterns in F_2s derived from crosses between cultivated rice and the weed red rice. They found that Mendelian expectations were met in only about 40 percent of the populations studied. Such segregation distortion is frequently observed for lineages that emerge from interspecies hybridization events involving nontransgenic parents (e.g., Zamir and Tadmor 1986). Interestingly, with very rare exceptions, transgenic traits in plants are almost universally more or less dominant (but see Halfhill et al. 2001 for a transgene allele, green fluorescent protein from jellyfish, expressed with incomplete dominance in plants). In contrast, it appears that the majority (though by no means all) of the traits that distinguish cultivated plants from their wild relatives are determined by recessive alleles at individual loci with major effect, sometimes modified by a few extra loci of minor effect (Gepts 2001). (A recent genetic analysis of domesticated traits in sunflower [Burke et al. 2002] found a different trend. Additive traits were the most common class; dominant and recessive traits were about in equal frequency.) A classic example is the locus for bitterness in the cucumber family. The wild relatives of cultivated cucumbers and other squashes are characterized by a dominant allele or alleles responsible for extremely bitter compounds (cucurbitacins) that render them so bitter as to be unpalatable; the cultivars are homozygous for the recessive allele and therefore edible (Barham 1953, Bates et al. 1990). That contrast is probably the biggest difference between transgene expression and the expression of other genes that separate crops from their wild relatives. It is not clear what, if any, the trend would be for traits incorporated into crops subsequent to domestication (P. Gepts, personal communication).

What is the significance of dominant crop traits in the fate of hybrids and hybrid derivatives? If the transgenic crop allele is dominant to its counterpart in the wild population (which is always the case for transgenes because the "allele" in wild plants is simply an empty location on the chromosome corresponding to the point of insertion in the transgenic plant—a situation called "hemizygosity"), then the transgenic hybrids will always express the trait created by the transgene. As discussed earlier, this appears to be the case for all transgenic hybrids studied. Recessive alleles are not expressed in heterozygotes. If traits are additive, then the hybrids will be more or less intermediate to their parents. Thus, in hybrids, dominant alleles are instantly subject to selection; recessive alleles are not. If transgenes are detrimental or beneficial in the wild, their disadvantage or advantage will be instantly expressed. But in hybrids, recessive alleles will act as neutral alleles.

The population-genetic consequences of advanced introgression are a simple but profound extension of the hybrid case. If hybrids are rare, then immigrant alleles will be rare as well. Alleles that are rare occur almost exclusively in the heterozygous condition. If they are recessive, in heterozygotes they will continue to act as neutral alleles, their fitness consequences as homozygotes notwithstanding. If rare immigrant alleles are dominant, selection will act immediately on them.

Consider the following example. Two types of crop alleles, A and B, have the potential to confer a spectacular fitness boost to an individual in a wild population. Allele A is recessive. Allele B is dominant. Both have introgressed into a wild population, both occurring at a frequency of 10 percent. Assume that the populations are randomly mating. On average, in a population of one hundred individuals, one individual would express the advantage associated with allele A, whereas nineteen individuals would express the advantage of allele B—almost a twenty-fold difference. In other words, natural selection can act much more effectively on dominant alleles. The take-home message is that the majority of crop alleles incorporated by traditional methods will tend to be masked from natural selection when they introgress into natural populations, while transgenes will be subject to natural selection from the start and thereafter.

What about the phenotypes expressed by the transgenes? The majority of transgenic phenotypes that have been created and field-tested can be crudely categorized into one of three groups. Initially plants were engineered with agronomic traits, with the goal of directly or indirectly increasing their yield, particularly with pest resistance and herbicide resistance. The vast majority of the cultivars that have been deregulated for commercial production fall into this category; in the United States, thirty-nine out of the fifty-three deregulated transgenic varieties (74 percent) had either pest resistance, herbicide resistance, or both, according to the ISB website (http://www.nbiap.vt.edu), as of late September 2001. Engineered genes are also used for quality traits, such as altered vegetable oil composition for a longer shelf life. The first transgenic variety deregulated in the United States was of this type, a tomato with delayed ripening (Kramer and Redenbaugh 1994). The third and newest category of engineered traits is that of plants serving as biochemical factories, producing pharmaceutical and other commercial biochemicals. One example is the commercial production of avidin, a compound used in biomedical research, by transgenic maize (Hood et al. 1997). The first two types of phenotypes, agronomic and quality-enhancing traits, are similar to those routinely created by other plant improvement methods. It may be that the introduced genes come from very different sources, but the primary environmental impacts of those transgenes, if they introgress into wild populations, are probably going to be more or less predictable. The third category is one that is quite different, but little has been written about the environmental impacts of such plants or their hybrids with wild relatives.

Will transgenes affect the likelihood of successful hybridization and subsequent introgression with wild relatives? As a group, probably not. A small but growing number of experimental studies have shown that transgenic crops are capable of spontaneously mating with wild relatives—and at rates on the order of what would be expected for nontransgenic plants (e.g., Jørgensen et al. 1998, Lefol et al. 1996b). Furthermore, a growing body of experimental work has revealed that the presence of a transgene does not in itself appear to be generally beneficial or detrimental in hybrids or hybrid derivatives (e.g., Mikkelsen et al. 1996, Snow et al. 1999, Spencer and Snow 2001), but see Linder (1998) for an example of detrimental effects of transgenes in hybrids and Oard et al. (2000) for beneficial effects. Of course, if transgenic-wild hybrids were exposed to the specific stress for which the transgene was chosen to provide protection, we would expect those hybrids to enjoy an advantage compared with nontransgenic hybrids. For example, resistance to a given herbicide would provide an advantage in the presence of that herbicide.

Still, in certain cases transgenic plants might be more or less successful in hybridizing with wild relatives than their nontransgenic counterparts. In some cases altered mating patterns might be the direct result of an intentional phenotypic change. For example, alfalfa, tobacco, and many other species have been genetically engineered so that plants produce little or no viable pollen, that is, they are male-sterile (Li et al. 1997, Rosellini et al. 2001). Obviously, plants that have full male-sterility will be unable to fertilize other plants. Of course, if they are female-fertile they can still engage in unilateral hybridization. Several engineered constructs have been proposed for the purposes of reducing hybridization or preventing the production of fertile hybrid offspring altogether (e.g., Gressel 1999). These strategies will be detailed in the next chapter. Also, one can imagine certain intentional traits, such as altered flower color or shape, that might end up unintentionally resulting in changing hybridization rates.

In fact, there is evidence that, as an *unintended* transgenic trait, increased flower size results in higher levels of pollen export in natural conditions. Bergelson et al. (1998) noticed that two *Arabidopsis thaliana* lines that were transgenic for herbicide resistance had slightly larger flowers than nontransgenic mutants for the same phenotype. The authors compared the outcrossing rates and found that the transgenic strains had about a twenty-fold increase in outcrossing rate relative to the mutants, whose outcrossing rate was typical for the species. Such unanticipated effects of an allele, called "pleiotropy," are not unusual for natural genetic variants (e.g., Hilu 1983), and we might expect pleiotropy to occur at about the same frequency for engineered alleles. Pleiotropic effects of transgenes might act to reduce introgression rates just as frequently as they act to increase them. An example was given in the previous chapter: hybrids of Swiss chard and transgenic sugar beet for virus resistance unexpectedly bolted significantly less frequently than their nontransgenic counterparts (Bartsch et al. 2001). The result of such reduced or delayed flowering would be to slow the introgression of transgenes.

Finally, altered hybridization rates might result from the geographic position of transgenic plantations relative to traditional varieties. For example, traits for tolerance of harsh environmental conditions are currently being engineered into certain crops (e.g., salinity; Winicov 2000, Roy and Wu 2001). Such traits may permit the crop to be grown in new regions, including sites adjacent to a wild relative from which it was previously geographically isolated by tens or hundreds of kilometers. Sometimes simply a change in distance of that magnitude is sufficient to allow natural hybridization between distinct but compatible species that had never previously hybridized (Levin 1978, Grant 1981).

Transgenic plants are a heterogeneous group. And they are going to become even more diverse. The tools of plant biotechnology—as well as the increasingly sophisticated tools of plant improvement without engineering—open up an unimaginable array of potential phenotypes and uses. It's not unreasonable to predict that computer chips will be growing on plants long before the end of this century.

However, it is clear from the foregoing discussion that *certain* current transgenic plants possess *certain* substantial differences from the current products of alternative technologies. We can use that information to anticipate whether and when certain transgenic crops may pose risks more frequently than their nonengineered counterparts. Let's return to the four concerns frequently expressed about the introgression of transgenes into natural populations.

Transgene Introgression and Weed Evolution?

The first such concern is that introgressed transgenes will result in the evolution of increased weediness in terms of a new weed, a more difficult weed, or an invasive species that threatens natural ecosystems. This possibility was among the earliest environmental concerns aired about the field release of transgenic plants (Colwell et al. 1985, Goodman and Newell 1985). Simply stated, the worry is that an introgressed transgene or set of stacked transgenes might confer a sufficient advantage to a wild plant that it evolves into a difficult weed (or, in the case of a wild relative that is already weedy, a more difficult weed). As discussed in chapter 9, introgression of alleles from nontransgenic cultivated plants has already, on occasion, resulted in such problem plants.

Should we expect transgenics to be any different? At the moment, most of the world's transgenic acreage is planted in genotypes engineered with alleles that, if they occurred in wild populations, might confer an advantage under the appropriate circumstances. About three-quarters of the transgenic crops deregulated in the United States are pest resistant and/or herbicide resistant. Furthermore, a small but growing number of transgenic plants are being field-tested with traits that confer tolerance to abiotic stresses, such as the ability to grow in soils with high salinity levels (e.g., Winicov 2000, Roy and Wu 2001). At first glance the new generation of plants engineered to produce commercial biochemicals might appear to offer little risk. However, avidin, the biomedical compound already commercially produced by transgenic plants in the field, has been shown to have insecticidal properties, endowing the plants that create it with a level of additional pest resistance (Kramer et al. 2000). (Fortunately, plants transformed to produce avidin have extremely high levels of male-sterility, reducing opportunities for transgene escape; National Academy of Sciences 2002.) Other traits, engineered for purposes other than to benefit the survival and yield of a crop, might also indirectly prove advantageous in the field, whether under cultivation or in the wild. The dominance of advantageous transgenic traits may help them enter and spread through wild populations. But, as discussed in chapter 9, an advantageous allele might simply spread through the population without altering the aggressiveness of the species. In many cases it will simply be difficult to judge a priori whether a transgenic phenotype will have a special advantage relative to its nontransgenic counterpart—and, if an advantage exists, whether it will translate into increased weediness or increased invasiveness.

However, herbicide resistance is one phenotype that deserves more discussion. Although crops with herbicide resistance can be created by traditional methods, the phenotype is a current favorite for engineered crops. In the United States, as of late September 2001, almost half of the deregulated transgenic varieties, twenty-three out of fifty three, are listed as having an herbicide-resistant phenotype, according to the ISB website (http://www. nbiap.vt.edu). But that's not the full story. Genes for herbicide resistance are sometimes also used as selectable markers during the engineering process to screen for transformed cells. For example, a transgene for glyphosate resistance was used as a selectable marker in the development of a potato developed for transgenic resistance to the Colorado potato beetle and to a viral disease (Petition #99-173-01). Thus, there are eight additional crops deregulated for commercialization of other phenotypes that are functionally herbicide resistant, bringing the total to thirty-one out of fifty-three, well over half.

Introgression of a crop transgenic for herbicide resistance into populations of a related weed that has been controlled by that herbicide isn't the end of the world, but it is certainly a problem. Because the transgene is dominant and the selective pressure from the herbicide will be strong, introgression should proceed rapidly, unless the hybrids are highly sterile. At best, the evolution of new herbicide resistance in a weed is a nuisance for farmers, who must find an alternative herbicide or method to control a previously controlled weed, and a source of lost revenue for the maker of the herbicide in question. At worst, herbicide resistance in the wrong plant (perhaps stacked with other advantageous transgenes) could spell the obsolescence of a popular, environmentally friendly herbicide, resulting in the loss of millions of dollars of potential profits and environmental damage from an environmentally unfriendly alternative.

As I write, there are still no reports of transgenic herbicide resistance naturally introgressed into wild plant populations. However, a recent incident in Alberta involving gene flow among herbicide-resistant crop varieties may give a taste of things to come. Spontaneous hybridization occurred among three different varieties (two transgenic, one not) of canola that had been planted near one another in 1997 and 1998, each resistant to a different herbicide (glufosinate, imidazolinone, and glyphosate) (Hall et al. 2000). As a result of the hybridization, by late 1998, volunteers were resistant to multiple herbicides. The resistance alleles moved rapidly. Scientists studying the volunteers reported that "a single triple-resistant individual was located more than 550 m from the putative pollen source 17 months after seeding" (Hall et al. 2000). Even though multiple-resistant *Brassica napus* volunteers are typically in low frequency, their presence warrants more complicated weed management. Websites from two different Canadian provinces have been created to provide advice for canola producers on how to avoid or manage outcrossing (e.g., http://www.agric.gov.ab.ca/crops/canola/outcrossing.html).

According to Linda Hall, the weed scientist who has studied the volunteers, the evolution of multiple resistance in the Prairie Provinces has not resulted in a major problem for most canola farmers growing seed to be processed into oil (personal communication). However, the potential of natural crossing between varieties and between varieties and volunteers in this region may preclude the production of seed of sufficient purity for the organic market and for seed growers. Hall also reports, "Weed control in canola has improved with the introduction of herbicide resistant crops. But they have increased the complexity of weed control in the rotation. Multiple resistant canola must be assumed to be present, limiting herbicide choices. On balance, weed control has become more dependent on information, record keeping and knowledge of herbicide groups. Many producers currently farming smaller acreages are not able to change quickly enough."

Transgene Introgression as an Extinction Risk?

The second concern is that transgene flow will increase wild relatives' chances of extinction. As illustrated in chapter 9, hybridization with non-transgenic species has already, on occasion, resulted in such problems for wild populations. Because extinction by hybridization varies primarily with the hybridization rate, only those transgenics that have enhanced hybridization with wild relatives will pose more of a threat than crops improved by other methods. As noted above, unless a transgene produces a phenotype that exports more pollen or otherwise enhances outcrossing (e.g., Bergelson et al. 1998, Muir and Howard 2001), the most likely way that hybridization with a transgenic will increase the extinction risk of a wild relative would be if the transgene alters the ecological tolerance of the crop so that it can now be grown in the vicinity of a previously isolated wild relative.

Helianthus paradoxus is a threatened species and is compatible with cultivated sunflower, *H. annuus,* but hybridization with that species is limited by the fact that *H. paradoxus* is unusually tolerant of brackish conditions (Rieseberg 1991b, Rogers et al. 1982). If a new allele for salt tolerance made it possible and desirable to grow cultivated sunflower in sizable plantations adjacent to the remaining stands of *H. paradoxus,* the populations of the rare species are so small that they would suffer immediately from the gene flow. Of the current transgenic phenotypes being produced and developed, it appears that only such niche-enhancing phenotypes are more likely to create problems than the typical product of traditional crop improvement.

Transgene Introgression That Depletes Genetic Diversity?

The two other concerns about transgene flow have been articulated much less frequently than the first two. The first of these is that transgene flow might cause the erosion of genetic diversity in wild populations. The loss of genetic diversity is a potential problem in two ways. In situ genetic diversity is often considered a germplasm resource for future crop improvement (e.g., Ladizinsky 1989). Also, some conservation scientists see standing genetic diversity in wild populations as insurance against extinction because it can be a substrate for natural selection, affording evolutionary flexibility under future environmental change (e.g., Lande 1988).

If an immigrant allele is new to a population, as we would expect for all transgenic alleles relative to the genetic composition of wild populations, genetic diversity should first *increase* with the initiation of transgene flow (table 4.3). In the case of transgene flow, the population will then be polymorphic for the presence or absence of the transgene. With regard to the transgene's locus, diversity cannot decrease relative to the prehybridization situation because, at the most extreme, the transgene will sweep through the population, replacing the effective "null" allele with the transgene allele. If one fixed allele is replaced by another, the net change in diversity is zero. Therefore, the introgression of a transgene in itself cannot directly lead to a decrease in genetic diversity.

But what about alleles linked to the transgene? Let's assume that the allelic diversity at those linked, nontransgenic loci is less diverse in the crop than in the wild population (a common assumption [see chap. 4] but not always the case [Blancas 2001]). If that is the case, the most substantial loss of genetic diversity due to a transgene will occur for tightly linked loci when a very beneficial transgene rapidly sweeps through a wild population (see the discussion of "selective sweep" in chap. 4). The fact that the transgene is dominant will facilitate that sweep. Alleles that are neutral or advantageous and are tightly linked to the transgene will also sweep through the population, replacing, and thereby reducing, any native diversity at those loci. The greatest impact will occur for those species that reproduce mostly by selfing (e.g., soybeans, barley) or asexual reproduction (e.g., sugarcane, cassava), which tend to maintain linkage; the least impact will occur for those species that reproduce mostly by outcrossing (e.g., maize, rye), which tends to break down linkage groups.

However, a transgene and tightly linked loci comprise a tiny part of a plant's genome, which contains on the order of twenty thousand loci (Meinke et al. 1998). Even if a plant contains several transgenes, their impact on local diversity will be small compared with the impact from the nontransgenic part of the genome. Gene flow from a transgenic variety will jeopardize local whole-genome diversity more than a corresponding nontransgenic variety only if (1) the genome-wide genetic diversity of the transgenic cultivar is substantially less diverse than the cultivar it replaces or (2) wild populations receive more gene flow from the transgenic cultivar than from prior cultivars of that species (as discussed in the preceding section).

Transgene Introgression as "Genetic Pollution"?

The final concern stated about transgene flow to wild populations is that the transgenes themselves may represent a source of pollution. The specific concern is that transgenes, as man-made constructs, represent "unnatural" genetic material. But how unnatural are transgenes? Transgenes that are simply one or more genes that are isolated and reintroduced into the same species are essentially no different from duplicated genes. Gene duplication is a common natural phenomenon, resulting from unequal crossing over, polyploidy, and other cytogenetic phenomena (Gottlieb 1982, Lynch and Conery 2000). Of course, if transgenes are introduced from related species with some degree, however small, of sexual compatibility, then the inserted transgene is essentially equivalent to a product of interspecific hybridization, which can be obtained by intentional breeding or natural hybridization.

However, most transgenic plants created to date have transgenes from organisms that are just about as distantly related as can be. While it would be impossible for traditional breeding to transfer genes between the kingdoms of life, interkingdom gene transfer does occur naturally (Syvanen and Kado 2002). Such "horizontal transfer" is defined as the nonsexual transfer of genetic material from one organism into the genome of another. For example, plants occasionally acquire genes from other kingdoms of organisms, such as fungi (e.g., Adams et al. 1998). The specific mechanisms for such transfer are poorly understood. The rate of acquisition is extremely low compared with gene transfer by natural hybridization between closely related species but is surprisingly high over evolutionary time scales. For example, flowering plants have acquired a certain mitochondrial gene by hundreds of independent horizontal transfer events over the past ten million years (Palmer et al. 2000). In essence, horizontal transfer is natural genetic engineering.

Plants with transgenes that are wholly artificial are still a tiny fraction of those so far created. Such transgenes may represent a class of genetic material that might never have evolved on its own. If that is indeed the case, then the presence of those genes in natural populations might be the conceptual equivalent of man-made chemicals in the environment. Still, I am uncertain about classifying the presence of transgenes in natural population as a hazard in itself. Whether artificial DNA that is otherwise benign poses a problem when it occurs in a natural population is a question of values. It is an issue better tackled by experts in philosophy, such as bioethicists.

Nontarget Impacts from Introgressed Transgenes?

An often discussed concern regarding the field release of transgenic plants is that it may have unintended impacts on organisms in the surrounding environment, affecting ecosystem structure and function (e.g., Keeler and Turner 1990, National Academy of Sciences 2000, Rissler and Mellon 1996, Scientists' Working Group on Biosafety 1998, Snow and Moran-Palma 1997, Tiedje et al. 1989, Wolfenbarger and Phifer 2000). For example, maize transgenic for a lepidopteran-specific insecticidal compound to effect resistance to a moth species might create and disperse insecticidal pollen that is toxic to other

Lepidoptera, with complex consequences for local ecosystems (Obrycki et al. 2001).

Presumably transgenes should produce the same nontarget impacts whether they are expressed in cultivated plants or in wild ones. Surprisingly, discussion of nontarget impacts from introgressed transgenes is extremely rare (but see Stewart 1999). The concerns are straightforward but hard to evaluate. Would the presence of transgenes in natural populations alter community structure and function? And more important, would transgenes pose more of a problem than other crop alleles that have been introgressing into natural populations for years?

Let's concentrate on the primary groups of transgenic phenotypes and see what impacts they may have when they reside in wild populations, keeping in mind that all ought to be expressed as dominant alleles. It is unlikely that an introgressed allele for herbicide resistance will have any immediate community effects in the wild. On the other hand, introgression of a transgene producing an insecticidal compound for insect resistance is likely to have instant effects because it would alter the community of animals that feed on the wild hybrids. If the compound persists in the soil before degrading (Stotzky 2002), it could also have effects on the soil ecosystem. Likewise, an allele that creates a commercial biochemical is likely to create the same biochemical in the wild. If that compound is toxic to members of the local community, then it too might alter ecosystem structure and function. It is not clear whether transgenes created to alter product quality (such as ripening time or seed oil composition) are likely to have a strong effect on ecosystems, although one could imagine that a trait as simple as altered seed oil composition could affect the nutrition of rodents and other animals that eat those seeds. Of course, an allele need not be transgenic in order to have nontarget effects in wild populations. Whether nontarget effects of transgenes in the wild would be more severe than the effects of nontransgenic crop alleles is worthy of carefully directed experimental research. Without data comparing the ecosystem impacts of pairs of transgenic and nontransgenic crop-wild hybrids, it is hard at this point to make any generalization.

It would be even more difficult to make predictions about the secondary nontarget effects of transgenic traits in natural ecosystems. For example, resistance-management planting schemes have been adopted for certain insect-resistant transgenic crops to slow the evolution of insects with resistance to the toxins produced by those crops (Mellon and Rissler 1998). No one will be managing wild populations bearing transgenes to slow the evolution of resistance in the insects that consume them. Indeed, they might be the same insects that eat the related crop. Furthermore, pleiotropic effects might have significant consequences in natural ecosystems. Saxena and Stotzky (2001) recently discovered that the cells of transgenic insect-resistant maize are highly lignified relative to nontransgenic maize, resulting in plants that decay more slowly when they die. While this pleiotropic effect may have limited consequences in agroecosystems, if expression is the same in wild maize relatives, the consequences may be different in nature, at the very least altering the physical structure of the litter layer of the ecosystem.

Because of the manifold interconnections between any species and other members of its ecosystem, it would be easy to speculate on nontarget impacts of introgressed transgenes. The few examples I have provided for this previously ignored topic are sufficient speculation for a population geneticist. I leave further discussion for community ecologists or, better yet, those scientists who have gathered experimental data comparing the nontarget impacts of transgenic and nontransgenic hybrids.

Some More Complex Considerations

Once transgenes enter wild populations, they have the opportunity to spread to other populations and to recombine. Just as genes can move from a domesticate into a wild taxon, there's no reason to assume that they will necessarily stop there. As a hypothetical example, one can imagine a transgene from canola introgressing into wild *Brassica campestris*, providing some advantage and sweeping through that species. Because *B. campestris* is widespread and cross-compatible with certain other *Brassica* weeds, it could then serve as a genetic "bridge," delivering that transgene to those species, possibly helping one or more of them to evolve to become more aggressive, even if the gene doesn't alter the aggressiveness of *B. campestris* itself.

Now consider the case of canola engineered to create a nonedible commercial biochemical. Introgressed *B. campestris* could serve as a genetic "reservoir" that could hold and return the transgene to a canola crop in a different place and time. This reverse introgression could be especially problematic if the recipient variety was one grown for human consumption and managed as an open-pollinated crop (i.e., the farmer saved and replanted the seed each year). If the transgene in question increased in frequency to the point that the concentration of the commercial biochemical reached toxic levels in the harvested seed, it could have human health effects. The scenario is very unlikely, but none of the steps is unrealistic

Finally, consider the possibility of natural "stacking" (Senior and Dale 1999). Just as genetic engineers can create varieties with multiple transgenes, so too might wild populations evolve that accumulate multiple transgenes by multiple hybridization events with several different transgenic varieties. So many kinds of transgenic maize have been created that one can imagine that if the Mexican government lifts its moratorium on transgenic maize, wild teosinte populations could "collect" alleles for resistance to different kinds of insects, for resistance to multiple herbicides, for the production of commercial biochemicals and pharmaceuticals, and for a host of other novel traits. Judging the potential environmental impacts of combinations of so many novel genes will be challenging.

Conclusions

Are the hybridization-mediated environmental impacts from transgenics likely to be different from those of traditionally improved crops? No, but that's not necessarily good news. As documented in previous chapters, gene flow from traditionally improved crops has on some occasions created problems by acting as a catalyst for weed evolution and by nudging wild relatives closer to extinction. We would expect the same for transgenic crops. It is also clear that *certain* current transgenic plants possess *certain* substantial differences from the current products of alternative technologies. Genetic engineering has the potential to create *certain* phenotypes that will be much more problematic than the average product of plant breeding, just as plant breeding has the potential to create *certain* phenotypes that will be much more problematic than the average product of genetic engineering. The good news is that the controversy about transgenic crops has stimulated some scientists to take a closer look at both traditional and novel methods for managing hybridization with the goal of reducing these problems. Gene-flow management is the focus of the next and final chapter.

CHAPTER 12

Whether and How to Manage Domesticated Gene Flow into Wild Populations

As a regulator working for the U.S. Department of Agriculture for 8 years evaluating proposals for field trials, I always assumed that if the probability of gene flow between a transgenic and a wild relative was not zero, then it was taken for risk assessment purposes to be one. Then and only then can one really focus on the relevant question, which is not "Will there be gene flow?" but "What consequences might be expected in the event of gene flow?"   —VAL GIDDINGS, VICE PRESIDENT, BIOTECHNOLOGY INDUSTRY ORGANIZATION, 2000

Treatises have been written on how to assess the risks of gene transfer. These discussions of the hazards and of risk assessment did not suggest that biotechnology could mitigate the risk. The governmental panels that regulate transgenic crops and those interested in regulation have not asked about ways to prevent weeds from using any traits that spread from crops. An analysis of crop and weed traits indicates both species-specific and general possibilities for mitigating gene transfer.

—JONATHAN GRESSEL, WEED SCIENTIST, 1999

Introgression from cultivated plants to their wild relatives sometimes creates problems in terms of the evolution of more difficult weeds, new invasives, and increased risk of extinction. The deployment of and controversy surrounding transgenic plants—as well as the recognition that the introduction of exotics may have undesirable consequences—has made most scientists planning to release novel plant phenotypes more mindful of the possible impacts. Note that I used the word "release" instead of "create." The problems of hybridization may come not only from the new genotypes of plant improvement but also from introduced exotic taxa (Ellstrand and Schierenbeck 2000). Scientists who release novel plants, transgenic or not, are the first line of gene-flow management. In many countries regulatory scientists make up the second line of management for specific classes of plants such as transgenic plants or exotic plant species that are deliberately introduced.

To Release or Not to Release?

The decision that both types of scientists make is the most important one for gene-flow management: whether or not to proceed with an introduction at all. I am aware of two decisions made by genetic engineers not to proceed with the development of transgenic products because of gene-flow concerns. Both decisions involved whether or not to create herbicide resistance in varieties that typically grow in close proximity to cross-compatible noxious weeds. Furthermore, in the mid-1990s an executive scientist in an agricultural biotechnology company told me that he had created a policy prohibiting engineering any plant species with cross-compatible relatives in the United States. Likewise, regulatory scientists often consider the potential consequences of gene flow in their determinations. For example, USDA-APHIS specifically mentions gene flow in its guidance to applicants seeking to field-test transgenic plants under permit and notification as well as to those petitioning for deregulation of those plants (APHIS 1997b). While the examples just discussed involve transgenic plants, choices can be made regarding any novel phenotype in which alleles might create a problem in wild populations.

If it is already known that the new phenotype will be grown close to (i.e., within a few kilometers of) a cross-compatible relative, deciding whether or not to proceed with a project for environmental reasons may prove challenging. It is impossible to devise a scheme to identify all possible consequences of every environmental impact of a new phenotype. But it is possible to outline the well-known impacts and provide guidance to create a mindful approach for examining consequences to the agroecosystem and adjacent ecosystems. At least two guides offer perspective on the environmental impacts of transgenic plants, and both cover the consequences of gene flow (Rissler and Mellon 1996, Scientists' Working Group on Biosafety 1998). The portions of both books addressing hybridization could apply equally well to novel exotic domesticated species and novel products created by traditional breeding.

More specific attention has been given to the two well-known risks associated with gene flow: weed evolution and extinction by hybridization. Weed evolution by hybridization depends on whether an introgressed allele or set of alleles confers traits that make a weed more difficult to control or alter its niche by removing one or more key constraints that limit that niche. While such evolution has been rare overall, a few of those cases have been real doozies (tables 9.2 and 9.3). While increased difficulty of weed control might be obtained simply by acquiring a single allele (e.g., one for herbicide resistance), a more complex set of factors may play a role. For example, certain novel traits (e.g., herbicide resistance, disease resistance, insect resistance) may enhance particular fitness components of a wild plant in certain environments but may or may not have attendant costs that may limit that fitness gain (Bergelson and Purrington 1996, Traynor and Westwood 1999).

Predictive discussions of gene flow as a stimulus for the evolution of increased aggressiveness focus almost exclusively on the spread of an advantageous allele or alleles (e.g., Ellstrand et al. 1999, Marvier and Kareiva 1999, Raybould and Gray 1994). Advantageous alleles alone may be necessary, but not always sufficient, to result in increased invasiveness (Bergelson 1994). For example, while it is clear that hybridization has played a role in the evolution of invasiveness in many species, it is not clear whether that invasiveness is primarily due to changes in fitness, changes in niche, or a combination of the two (Ellstrand and Schierenbeck 2000). Additionally, why many natural hybrids do not evolve invasiveness is a mystery. The ecological genetic basis of the evolution of invasiveness remains largely unexplored and represents a research field ripe for study. The bottom line is that while logic or experiment may clearly demonstrate that a new allele or set of new alleles provides a wild plant with an advantage relative to other plants, it may often be impossible to predict whether that advantage will translate into significantly increased aggressiveness that creates a new plant pest or makes an old pest worse.

One more avenue for obtaining information about the possibility of weediness would be to seek advice from weed science and invasion biology. A good literature abounds, both in print and online, on the current status of invasive and weedy plants. There are many experts on noxious plants who can provide valuable insights. Consulting the literature or such experts should give the plant breeder/genetic engineer some inkling of whether a transgenic hybrid or its descendants might carry a risk of creating a plant pest.

What about the risk of extinction by hybridization? This problem has received considerable theoretical attention. Predictive models have been developed that demonstrate whether and how quickly extinction occurs with a variety of parameters, parameters that can often be obtained from descriptive or experimental work (e.g., Huxel 1999, Muir and Howard 2001, Wolf et al. 2001). In general, hybridization rate is important for determining the speed of extinction. These models could prove useful for judging whether releasing a cultivated plant in closer proximity to a rare wild relative will increase the risk of extinction of that wild population. Likewise, they could be of value to a plant conservation scientist managing an endangered species already suffering from hybridization with a related domesticate nearby.

A case study in one of these papers (Wolf et al. 2001) actually considered "real-life" data involving gene flow from a crop, cultivated sunflower (*Helianthus annuus*), into a wild population of the same species. Based on the authors' research experience with sunflowers, certain assumptions were made: "Cultivar population size remained constant and contributed pollen but not seeds to the [wild] population under study. . . . The number of hybrid and wild plants each generation, however, was controlled by reproduction and competition. . . . In an additional simulation, . . . cultivated plants were present for only one breeding season" (Wolf et al. 2001). The results were dramatic and varied with initial conditions (note that the authors use the term "hybrid" for any plant with hybrid ancestry):

> When the cultivar was sown for only a single generation, the fate of wild *H. annuus* depended on the initial frequencies of wild and cultivated plants. If the initial population had equal numbers of each parental type ($N_1 = N_2 = 100$), the wild type persisted and hybrids were eliminated within a few generations. . . . But when the wild population was much smaller than the number of cultivated plants ($N_1 = 100$, $N_2 = 10,000$), as is likely to be the case in nature, the wild plants were completely replaced by hybrids 75% of the time, with a mean time to extinction of 4.0 ± 2.1 generations.
>
> When the cultivar was present every generation, wild plants were replaced by hybrids under much less stringent conditions, including equal initial population sizes. For example, when the competitive abilities of hybrids and wild plants were

equal, hybrids replaced wild plants in 18.1 ± 4.5 generations. Even when hybrids had a slight competitive disadvantage, . . . they were able to replace wild plants in 11.5 ± 2.1 generations if the cultivars initially had a numerical advantage (N_1 = 100, N_2 = 200). (Wolf et al. 2001)

The authors observe that their results are in accord with empirical work showing the rapid spread and persistence of cultivar alleles in natural populations of *H. annuus* (cf. "#12. Sunflower Seed" in chap. 7).

How to Manage?

Let's assume that the decision to release a new plant has been made (and that regulatory approval, if necessary, has been given) but that lingering worries about hybridization are sufficient to warrant some measures to minimize gene flow. Indeed, transgenic plantations of products under permit and notification in the United States are regulated by USDA-APHIS, whose regulations require conditions that limit gene flow by pollen, seed, and other reproductive propagules (APHIS 1997b). There is a misconception that permits and notifications are used only for small-scale experiments during the research and development stage of novel transgenic plants (cf. Kareiva and Marvier 2000). A number of commercial products, particularly biochemicals, are created by plants grown under permit and notification (National Academy of Sciences 2002). These plants are required by law to be reproductively isolated from other cultivars and cross-compatible wild relatives. Therefore, gene flow must be managed. Likewise, conservation managers may have little choice about the plantation of a crop near their reserves and so may need to consider management options.

Before examining the options, it is worthwhile to consider briefly how low gene-flow levels must be to be effectively nil for biological purposes. The algorithms described in chapter 4 or the models mentioned above may provide the gene-flow manager with an order of magnitude at or under which a certain problem will be negligible. In the face of a substantial concern, a compelling need to release the crop, and the absence of any other information, a good rule of thumb might be that acceptable gene flow is on the order of magnitude of the typical mutation rate (cf., Gressel 1999), that is, 10^{-6} to 10^{-7} (e.g., Grant 1975, Mange and Mange 1990). Note that spontaneous hybridization rates, depending on the species involved, can range from considerably lower than this level to considerably higher (table 4.1). Thus, in some cases gene-flow management may be unnecessary. In other cases it will be a requirement.

The "Starlink" incident of 2000 illustrates the importance of effective containment. This particular variety of maize was released for animal consumption prior to determination of whether it was approved for human consumption. Nonetheless, it rapidly entered the general maize grain supply of the United States, within a year turning up "in nearly one-tenth of 110,000 grain tests performed by U.S. federal inspectors" (Haslberger 2001). While it is highly unlikely that the Cry9C protein produced by this variety will prove to pose any kind of risk, the fact that pollen and seed moved the gene so rapidly demonstrates how quickly unintentional movement can occur. And because alleles of concern, transgenic or not, are multiplied by sexual reproduction, once unintentional spread occurs, recall may become extremely difficult, if not impossible.

What are the options? Two broad categories for gene-flow management have been proposed, mostly in the context of preventing transgene escape. These are ecologically based methods and genetically based methods. Ecologically based methods are those that depend on manipulation of ecological parameters, such as time and space, to reduce gene flow. They are cultural methods practiced by the grower of the plant rather than by the scientist who releases it. They are the only set of options usually available to the conservation scientist managing an endangered crop relative. The proposed genetically based methods are based on genetic manipulation of the plant prior to its release, such that gene flow by pollen or introgression is reduced. Interestingly, mechanisms that are effective in preventing the escape of transgenes into natural populations will probably prove equally effective in preventing the escape of intellectual property, a topic of great concern for commercial developers of transgenic plants (e.g., Ellstrand 1989, Crouch 1998). I consider each category separately.

Ecological Management of Gene Flow

Breeders and others concerned with the production of seed that is of a high level of genetic purity have long been concerned about ecological management of gene flow to minimize what they call "pollen contamination." Pollen contamination would be defined by seed producers as the unintended pollination of plants in seed production fields by other genotypes growing outside those fields, resulting in undesirable "off-type" progeny. For instance, it wouldn't do for the seed of long white "daikon"-type radishes to be infiltrated by the red spherical genotypes. Pollen contamination is such a bane to breeders and seed companies that various strategies have been developed to assure that a certain fraction of seed produced (e.g., 99 percent) are genetically pure (Kernick 1961). These strategies could be applied for reducing crop-to-wild pollen flow but, as we shall see, are unlikely ever to result in complete containment. Furthermore, keep in mind that cultural methods to restrict pollen contamination have been developed with the idea of keeping pollen out of a single plantation, as opposed to the multilateral pollen exchange between many crop populations and many wild populations.

Isolation Distances

One obvious strategy is to make sure that crops are planted at a certain distance from their wild relations. "Isolation distances" have been developed from experimental and anecdotal data for hundreds of crops (Kernick 1961, Kelly and George 1998). For outcrossing crops, isolation distances designed to attain 99 percent or greater levels of genetic purity typically range from hundreds of meters (e.g., for maize) to over a thousand meters (e.g., for beet); for crops that primarily self-fertilize, isolation distances can be as low as zero (e.g., for wheat) (Kernick 1961, Kelly and George 1998). Note that isolation distances are generally considered to be tools to reduce, but not eliminate, gene flow by pollen.

Such distances are more rules of thumb than hard scientific facts. That being the case, it is not surprising that they vary, depending on, for example, the genotypes involved or the region in which the crop is planted (e.g., maize) (Kernick 1961, Kelly and George 1998). Experiments measuring spontaneous hybridization rates between crops and their wild relatives in the field show a general decrease with increasing distance (examples in table 4.1). But they also show a considerable amount of variation, between years, sites, and replicates within sites. Klinger and Ellstrand (1999) reviewed a number of ecological factors that are likely to contribute to that variation.

For example, the relative sizes of populations involved in bilateral gene flow are often an important determinant of gene-flow rates. Experiments have often shown that gene flow is asymmetric between populations of very different sizes such that smaller populations receive more gene flow from large populations than large populations do from small ones (e.g., Handel 1983). We would expect that the necessary isolation distances should vary with the relative sizes of the source and sink populations.

Isolation distances are usually developed to prevent pollen contamination of one cultivar by another cultivar, but the dynamics of cross-fertilization may be different when a cultivated species is paired with a wild relative. While crops and distantly related wild plants often have relatively higher levels of reproductive isolation than pairs of closely related cultivars, this may not always be the case. For example, under normal growing conditions, natural rates of intercrossing between highly self-fertilizing cultivars of wheat may be very close to zero (Waines and Hegde 2002), but spontaneous hybridization between wheat and certain related weeds in the genus *Aegilops* is generally higher (Hegde and Waines 2002).

Relying on farmers to maintain isolation distances of hundreds of meters between outcrossing crops and wild relatives may involve substantial effort and expense if the wild relative is a common weed. I know of one case in which a company field-testing a transgenic plant refused a suggestion by a public interest group to extend the proposed isolation distance around the test from two hundred meters to four hundred meters. Indeed, assuming a field test measuring one hundred by one hundred meters, moving from two hundred to four hundred meters' isolation distance would more than triple the isolation area to be managed. Unless substantial isolation distances are easily obtained, for example, because of the natural distribution of wild relatives, isolation distance is a crude, costly, and probably ineffective way to manage gene-flow exchange between a domesticate and a compatible wild relative.

Surrounding a Field with Plants to Interfere with Pollen Flow

Scientists seeking to minimize pollen contamination have developed another cultural technique: surrounding their seed production fields with other plants to interfere with immigrating pollen that might otherwise fertilize their crop. Such plantations, referred to as "border rows" or "pollen trap crops" (Kelly and George 1998), are often used to enhance the effect of isolation distance in reducing pollen contamination. Scientists motivated to contain transgenes within a single field have conducted experiments to determine whether surrounding a field with a different species or a different genotype of the same species will effectively absorb pollen flow out of that field.

For example, sugar beet breeders plant a strip of hemp, *Cannabis sativa,* around their seed multiplication plots. These relatively tall plants with sticky leaves trap incoming pollen from other cultivars than might otherwise contaminate the seed being produced (Barocka 1985). Saeglitz et al. (2000) tested whether such a strip would be effective in capturing pollen moving in the other direction, exported from the sugar beet plot. The authors used thirty transgenic sugar beets as their pollen source. Around these they planted twelve male-sterile trap plants. Around those forty-two beets they planted a three-meter-wide barrier of hemp. Just outside this rectangular perimeter they planted twelve more trap plants as well as four additional trap plants in each of the eight primary compass directions at distances of up to three hundred meters. Seed set on the male-sterile plants served as an indicator of pollen receipt from the male-fertile transgenic plants. Genetic analysis of the seeds set on the trap plants confirmed paternity by the transgenics, as a way of checking for long-distance contamination from beets offsite from the experiment. No pollen contamination of the experiment was detected. Although trap plants varied considerably in seed set, the trap plants immediately inside the hemp strip and those immediately outside had roughly the same levels of pollen receipt. At greater distances, seed set generally decreased with distance from the pollen source, with the transgenic plants siring seed on plants as distant as two hundred meters. Plants three hundred meters from the pollen source set no seed. Clearly, the hemp strip is not effective at absolute containment; however, in this experiment, at least, it appears to reduce the isolation distance to considerably less than the eight hundred to sixteen hundred meters recommended for commercial seed production in the absence of other methods to limit pollen flow (Kernick 1961). In a similar experiment with transgenic canola, Morris et al. (1994) demonstrated that trap crops were more effective in reducing (but not eliminating) gene escape than barren zones (see also Staniland et al. 2000). Interestingly, experiments involving trap crops all appear to address how this strategy affects gene-flow patterns within the same crop. I am not aware of any experimental work that measures how the presence of trap crops affects gene flow from a cultivated species to a wild relative.

Other Ecological Management Options

Although other ecological management options are rarely discussed for gene-flow management, one particular method might prove more effective than either isolation distances or supplemental plantations to capture emigrating pollen. Isolation in time is typically more effective as a reproductive barrier than isolation in space (e.g., Levin 1978). For example, the hybridization rate between rice cultivars and the weed red rice decreases with increasing difference in their flowering periods (Langevin et al. 1990). If an annual crop has a short and very uniform flowering time, planting it a few days earlier or later than usual could alter spontaneous rates of hybridization with related weeds or endangered species. Conservation managers attempting to minimize gene flow into an endangered species might be able to persuade farmers who plant a cross-compatible crop nearby to modify their planting dates or to buy cultivars that have flowering times that do not overlap with that of the endangered species.

Genetic Management of Gene Flow

It is unreasonable to expect that those growing domesticated plants should bear the brunt of gene-flow management. The ecological methods just discussed may be hard for farmers to implement without suffering a reduction in crop yield or quality, especially if substantial acreage is set aside for crop isolation or for alternative crops of little value grown only to capture outgoing pollen. It is possible to imagine specific scenarios by which a farmer could benefit under such management. For example, planting a crop a bit early may reduce weed competition or reduce the probability of the attack of a certain pest. Still, if those scientists who are producing and advocating a crop are mindful that gene flow from that crop may create a problem, then it makes sense that they should be the ones to take responsibility for gene-flow management by creating products genetically designed to minimize that hybridization.

For example, Gressel (1999) has complained that everyone talks about transgene flow but no one has done anything about it. He suggests that there are plenty of genetic solutions for what he calls "failsafes" that limit transgene flow to wild relatives or for creating constructs that reduce transgene flow by putting hybrids at such a disadvantage that they cannot persist in the environment. Let's consider Gressel's (1999) three failsafes and others proposed for reducing introgression.

Failsafe #1. Apomixis

Apomictic plants reproduce by asexually produced seed and/or by vegetative propagation (Grant 1981, Richards 1997). If a plant is fully asexual and male-sterile, like certain potato cultivars, then it cannot transfer its alleles to other plants. However, many apomictic species retain low to moderate levels of sexual seed production (Asker and Jerling 1992, Grant 1981). Furthermore, moderate to high levels of pollen fertility are not rare for apomictic plants (Asker and Jerling 1992, Grant 1981). If that fertile pollen introduces alleles for apomixis into natural populations, those populations might suffer a unique problem. All other things being equal, alleles for apomixis have an unconditional advantage that could result in the uncontrollable spread of those alleles (Van Dijk and Van Damme 2000). If those alleles are linked to the transgene, the resulting "selective sweep" could spread the transgene much more quickly and effectively than a nonapomict. Except for a handful of specific crops that bear very little or no sexual seed and are highly male-sterile, apomixis is unlikely to be an easy solution for preventing the introgression of domesticated alleles into wild populations.

Failsafe #2. Chromosome-Specific Strategies for Allopolyploids

Many crops (e.g., bread wheat, peanut, coffee) are allopolyploids, housing multiple genomes derived from different sources. Often, only one of the genomes of the crop is compatible with that of a related weed, allowing alleles on those chromosomes to be transmitted to the weed by backcrossing with hybrids while alleles on the other genomes have a considerable transmission barrier. For example, bread wheat, *Triticum aestivum,* and the weed *Aegilops cylindrica* have the same "D" genome. However, bread wheat also contains two sets of chromosomes that are considered to be incompatible with those of *Ae. cylindrica,* "A" and "B." If a novel allele is incorporated into the chromosomes of bread wheat that make up either the "A" or "B" genome, it is predicted that they should have considerable difficulty introgressing into wild populations of *Ae. cylindrica* through backcrossing. Gressel (1999) predicts that "the risk of transgenic traits spreading into weeds can be reduced by orders of magnitude by using cytogenetic mapping to locate transgenes and releasing only those transgenic lines in which it is on a genome incompatible with local weeds." Traditional breeders could also take steps to put alleles of concern on an incompatible genome. I am not aware of data for the wheat-*Aegilops* system that show whether preferential exclusion of chromosomes is sufficient to qualify as a "failsafe" to reduce introgression by "orders of magnitude."

However, some data are available for a similar cytogenetic situation involving the polyploid crop oilseed rape. Oilseed rape, *Brassica napus,* shares one set

of chromosomes (the "A" set) with the weed *B. campestris* (= *B. rapa*). Therefore, it might be expected that if alleles of concern were placed on rape's other
set of chromosomes (the "C" set), they would be preferentially excluded in the
wild (Gressel 1999). When Metz et al. (1997) observed a strong decrease in the
frequency of a transgene in progeny resulting from a backcross of a *B. napus* ×
B. campestris hybrid to *B. campestris,* they explained that decrease as being a result of such preferential exclusion. Tomiuk et al. (2000) examined the situation more closely. First, they found that the cytogenetic data of others (Fantes
and Mackay 1978) showed no preferential exclusion of the "C" genome in
backcrosses. Second, they created a model to examine the data of Metz et al.
(1997) more closely. They found that an alternative, equally parsimonious,
hypothesis could not be excluded: that the "decrease in the frequency of
transgenic plants within the first backcross generation can also easily be explained by selection against transgenic *A*-chromosomes of *B. napus*" (Tomiuk
et al. 2000). They conclude, "Without more detailed genetic information, . . .
no decision can be made in favor of the *A*- or *C*-genome as the safer candidate
with respect to the introgression of transgenes into wild populations"
(Tomiuk et al. 2000). I am not aware of any data supporting the hypothesis
that placing alleles of concern on the appropriate cytogenetic location effectively reduces their introgression into wild populations. Clearly, there is a
need for experimental work of this type.

Failsafe #3. Alleles Specific to the Genomes of Organelles

In many plant species the chloroplast and mitochondrial genomes are inherited only maternally. If an allele of concern is engineered into the genomes
of either organelle of such a species, pollen cannot serve as a vehicle for its escape into wild populations (Gressel 1999). Unfortunately, seed can still serve
as such a vehicle. Consider the evolution of Europe's weed beet (chaps. 1 and
6) from the hybrid seeds initially borne by cultivated, not wild, plants.

Furthermore, contrary to popular belief, uniparental inheritance of
chloroplasts is not universal for plants (Smith 1989). In fact, an initial attempt
to contain transgenic herbicide resistance by engineering the chloroplast genome of tobacco (Daniell et al. 1998) drew considerable criticism from those
who pointed out that tobacco is one of the plant species in which chloroplasts
can be transmitted by pollen (e.g., Stewart and Prakash 1998).

In contrast, engineering the chloroplast genome of oilseed rape *(Brassica
napus)* might prove an effective way to slow or prevent pollen from carrying

alleles of concern from that crop into wild populations of the weed *B. campestris*. Scott and Wilkinson (1999) genetically analyzed forty-seven natural hybrids of the two species that were grown from seed from wild *B. campestris* plants. They found that the chloroplast genome of each hybrid matched that of its maternal parent. Although their sample size was small, it was sufficient to detect paternal inheritance occurring at levels of about 2 percent. But the authors' conclusion—that "there will be no or negligible pollen-mediated chloroplast dispersal from oilseed rape"—may be an overstatement. Their sample size was not sufficient to be a reliable estimator of paternal inheritance at the 1 percent level, which is still far higher than the desired level on the order of the mutation rate.

Other Failsafes

Genetic engineers are beginning to offer other solutions for containing crop genes, motivated by concern for both the environmental impacts of transgenes and the security of intellectual property. One example is the patented, but still hypothetical, "terminator" construct that, once activated, is designed to kill seeds just before they reach full maturity (Crouch 1998, Oliver et al. 1998); it is not clear how that construct would be expressed in pollen, that is, whether it would kill pollen or whether living pollen from activated plants would sire dead seeds on others. Other scientists are designing and experimenting with other constructs to kill or sterilize hybrids (e.g., Kuvshinov et al. 2001). One new technology seems particularly promising. Strategies have been developed to allow the excision of genes from transgenic plants (Hare and Chua 2002, Keenan and Stemmer 2002). Such constructs have already been effective at removing selectable markers from the genome of transgenic plants. Tissue-specific or chemically induced excision of transgenes from the genome of cells fated to produce pollen and seeds should prevent their transmission to future generations. Also, a number of species have been engineered for male sterility (e.g., Li et al. 1997, Rosellini et al. 2001). Another proposal is to identify alleles or loci that are responsible for physiological cross-incompatibility among species and to use those to reinforce isolation barriers between compatible or semicompatible lineages (Roseboro 2001). I am not aware of a construct that has been tested under a variety of field conditions and proven to be uniformly reliable and effective in eliminating or reducing bilateral hybridization. However, male sterility is one strategy that offers the protection of endangered species from the risk of extinction by hybridization, while most of the other options for genetic management appear to offer only some protection against weed evolution.

Each of the proposed failsafes may, in specific cases, be effective in slowing or preventing the introgression of alleles of concern from domesticated plants into natural populations. Obligate apomicts will never set hybrid seed. If a domesticated species has absolute maternal inheritance of chloroplasts, hybrids will never receive genes in those plastids from their paternal parent. But the efficacy of any of these proposed systems must be tested by experiments under field conditions. Presently, not one of the options provides a general panacea for bilateral containment of domesticated alleles.

Gene Flow Mitigation, beyond Failsafes

It is not clear that any of the proposed failsafes meets the dictionary definitions of "failsafe," that is, a measure "capable of compensating automatically and safely for a failure" or "guaranteed not to fail" (*American Heritage Dictionary of the English Language* 1992). More straightforward, and perhaps potentially more effective, are what Gressel (1999) proposes as "tandem constructs" for "transgenic mitigation (TM)." Here is the overall scheme:

> (1) Tandem constructs of genes act genetically as tightly linked genes, and their segregation from each other is exceedingly rare.
>
> (2) There are traits that are either neutral or positive for crops that would be deleterious to typical weeds, crops as volunteer weeds or wild species.
>
> (3) Because weeds compete strongly amongst themselves and have a large seed output, even mildly harmful traits will be eliminated from populations.
>
> Thus, if the gene of choice being engineered into a crop is flanked by TM genes in a tandem construct, the overall effect would be harmful if it spread into weeds. Even if one of the TM genes disappears, the other flanking TM gene will still provide mitigation. (Gressel 1999)

Gressel (1999) discusses at length traits that would be beneficial under cultivation but detrimental to plants in the wild. Traits that he advocates are those typically common in most modern agronomic crops: lack of secondary seed dormancy, uniform ripening, lack of shattering, dwarfing, and so on. He also lists alleles that suppress bolting or flowering for plants whose product is not a reproductive structure, like carrots and potatoes. Note that, for tandem constructs to be effective, the traits conferred by the flanking alleles must be dominant relative to their counterparts in wild plants. Although the genetic basis and chromosomal location for such traits is largely unknown for most crops at this point, data are beginning to accumulate (e.g., Burke et al. 2002, Gepts 2001).

Although Gressel (1999) proposes such genetic constructs to reduce the flow of transgenic alleles of concern from crops to wild relatives, they could be created for nontransgenic crops as well, albeit with more difficulty. With more thorough mapping and genomic analysis of important domesticated species, in the near future specific chromosomal locations could be targeted as sites for locating nontransgenic alleles of concern. Such alleles could be placed on chromosomal rearrangements, such as translocations or inversions, to shelter them from recombination. At the present, although they are still hypothetical, tandem constructs offer the best option for a general genetic solution to the problem of limiting gene flow to prevent the evolution of new or more difficult weeds and invasives. It is clear that genetic options for gene-flow management, if well designed and thoroughly tested, may provide a sufficient reduction in hybridization rate to allow the field release of plants bearing alleles of concern. Until those options are developed, decisions to release plants with alleles of concern must proceed with caution. Readers seeking more detail regarding molecular strategies for transgene containment are referred to Daniell's (2002) recent review on the topic.

Other Management Options

The motivation to prevent transgene flow may lead to solutions that are so creative that they do not fit into the two broad categories just discussed. Lev-Yadun and Sederoff (2001) suggested, for woody species, grafting nontransgenic scions onto transgenic rootstocks that have not yet reached reproductive age and that have had their branches pruned. Then the only reproductive structures that are formed on those plants are nontransgenic. In those species that have adventitious growth from the roots, these scientists suggest double grafting, starting with a transgenic shoot grafted to a wild-type rootstock, and subsequent grafting with a nontransgenic scion. Surely many new proposals for containment, both simple and complex, will be offered in the near future. But at the moment, Daniell (2002) best summarizes the situation: "As yet, however, no strategy has proved broadly applicable to all crop species, and a combination of approaches may prove most effective."

And in some cases the decision not to field-release but to grow a novel phenotype under physical containment may reap more than environmental benefits. If a plant is grown to produce a very valuable biochemical product, growing that plant in a greenhouse ensures not only a measure of environmental biosafety but also greater security of valuable intellectual property and a more uniform environment with better pest control, which should result in higher product quality. Indeed, as recently as March 2000, Val Giddings, vice president of the Biotechnology Industry Organization, in a comment written to an e-mail discussion group, predicted that such plants would be too valuable to be planted outdoors. It is surprising that any high-value industrial chemicals are commercially produced from field-grown, rather than greenhouse-grown, transgenic plants.

A Word about Monitoring

Monitoring for increased weediness or increased extinction risk due to gene flow might seem a more tractable option than trying to manage gene flow from the start. In the case of extinction by hybridization, measuring incoming gene flow by pollen with paternity analysis of seed set by a threatened species might be tractable if genetic markers are available to identify hybrid progeny. That parameter and other easily estimated parameters could be inserted into the preexisting theoretical models mentioned earlier in this chapter to determine whether hybridization from a domesticated species actually poses an extinction threat and, if so, whether that threat is urgent.

The situation is more complicated for weed evolution by hybridization. The evolution of increased invasiveness is not a simple function of hybridization rate or even an increase in fitness. Thus, measuring gene flow from a crop into a weed will rarely be sufficient to establish whether a new plant pest problem is pending.

Instead of measuring gene flow, the occurrence, fitness, and spread of hybrid derivatives could be monitored to determine whether those plants pose a new pest problem (Marvier et al. 1999). This is much more easily said than done. A striking lesson from invasion biology is that many introduced plants maintain small and innocuous populations for decades before erupting into invasiveness (Marvier et al. 1999). Thus, any program monitoring for significantly increased invasiveness must be planned for years. And that program's sample size and geographic extent must be large enough detect a weed problem early enough that eradication measures will be effective before it is too late (Marvier et al. 1999). Simulations with realistic parameters (Marvier et al. 1999) have shown that inadequate sampling can miss the initial invasion by more than ten years.

Some scientists have concluded that "containment of transgenic plants is not a viable option" (Kareiva and Marvier 2000), and that monitoring, followed by control measures, is a better option for identifying and correcting adverse environmental effects of transgenic crops (Kareiva et al. 1996, National Academy of Sciences 2002). But there will be circumstances in which monitoring will be inadequate or prohibitively expensive (Marvier et al. 1999). Monitoring is no substitute for containment. There may be some novel plants for which some hybridization will occur and the hazards will be clearly so great that environmental release should not occur. For others the hazards will be modest enough that release will be reasonable if one or more options are employed to substantially reduce the hybridization rate. In particular, the proposed genetic options are worthy of more refinement and experimental testing to determine their efficacy. An ounce of prevention is worth a pound of cure.

References

Abbott, R. 1992. Plant invasions, interspecific hybridization and the evolution of new plant taxa. *Trends in Ecology and Evolution* 7:401–405.

Abe, J., A. Hasegawa, H. Fukushi, T. Mikami, M. Ohara, and Y. Shimamoto. 1999. Introgression between wild and cultivated soybeans of Japan revealed by RFLP analysis for chloroplast DNAs. *Economic Botany* 53:285–291.

Acosta, J., P. Gepts, and D. G. Debouck. 1994. Observations on wild and weedy accessions of common beans in Oaxaca, Mexico. *Annual Report of the Bean Improvement Cooperative* 37:137–138.

Adams, K. L., M. J. Clements, and J. C. Vaughn. 1998. The *Peperomia* cox I group I intron: Timing of horizontal transfer and subsequent evolution of the intron. *Journal of Molecular Evolution* 46:689–696.

Adams, R. P., and B. L. Turner. 1970. Chemosystematic and numerical studies of natural populations of *Juniperus ashei* Bush. *Taxon* 19:728–750.

Ahmad, F., A. E. Slinkard, and F. F. Scoles. 1988. Investigations into the barrier(s) to interspecific hybridization between *Cicer arietinum* L. and eight other annual *Cicer* species. *Plant Breeding* 100:193–198.

Aldrich, P. R., and J. Doebley. 1992. Restriction fragment variation in the nuclear and chloroplast genomes of cultivated and wild *Sorghum bicolor*. *Theoretical and Applied Genetics* 85:293–302.

Aldrich, P. R., J. Doebley, K. F. Schertz, and A. Stec. 1992. Patterns of allozyme variation in cultivated and wild *Sorghum bicolor*. *Theoretical and Applied Genetics* 85:451–460.

Alexander, H. M., C. L. Cummings, L. Kahn, and A. A. Snow. 2001. Seed size variation and predation of seeds produced by wild and crop-wild sunflowers. *American Journal of Botany* 88:623–627.

Allard, R. W. 1999. History of plant population genetics. *Annual Review of Genetics* 33:1–27.

Allendorf, F. W. 1983. Isolation, gene flow, and genetic differentiation among populations. In *Genetics and conservation: A reference for managing wild animal and plant populations*, ed. C. M. Schonewald-Cox, S. M. Chambers, B. MacBryde, and W. L. Thomas, 51–65. Menlo Park: Benjamin-Cummings.

Allendorf, F. W., R. F. Leary, P. Spruell, and J. K. Wenburg. 2001. The problems with hybrids: Setting conservation guidelines. *Trends in Ecology and Evolution* 16:613–622.

Altieri, M. A. 2000. The ecological impacts of transgenic crops on agroecosystem health. *Ecosystem Health* 6:13–23.

Alvarez, M. L., S. Guelman, N. G. Halford, S. Lustig, M. I. Reggiardo, N. Ryabushkina, P. Shewry, J. Stein, and R. H. Vallejos. 2000. Silencing of HMW glutenins in transgenic wheat expressing extra HMW subunits. *Theoretical and Applied Genetics* 100:319–327.

The American Heritage dictionary of the English language. 1992. Boston: Houghton Mifflin.

Ammann K., Y. Jacot, and P. Rufener Al Mazyad. 1996. Field release of transgenic crops in Switzerland, an ecological risk assessment. In *Gentechnisch veränderte krankheits- und schädlingsresistente Nutzpflanzen. Eine Option für die Landwirtschaft?* ed. E. Schulte and O. Käppeli, 101–157. Bern: Schwerpunktprogramm Biotechnologie des Schweizerischen Nationalfonds.

Anderson, E. 1949. *Introgressive hybridization.* New York: Wiley.

Anderson, E. 1952. *Plants, man, and life.* Boston: Little, Brown, and Co.

Anderson, E., and G. L. Stebbins. 1954. Hybridization as an evolutionary stimulus. *Evolution* 8:378–388.

Anderson, M. K. 1997. From tillage to table: The indigenous cultivation of geophytes for food in California. *Journal of Ethnobiology* 17:149–169.

Anderson, R. L. 1993. Jointed goatgrass *(Aegilops cylindrica):* Ecology and interference in winter wheat. *Weed Science* 41:388–393.

Antonovics, J. 1968. Evolution in closely adjacent populations. V. Evolution of self-fertility. *Heredity* 23:219–238.

Antonovics, J. 1976. The nature of limits to natural selection. *Annals of the Missouri Botanical Garden* 63:224–247.

Antonovics, J., and A. D. Bradshaw. 1970. Evolution in closely adjacent populations. VIII. Clinal patterns at a mine boundary. *Heredity* 25:349–362.

Antonovics, J., N. C. Ellstrand, and R. N. Brandon. 1988. Genetic variation and environmental variation: Expectations and experiments. In *Plant evolutionary biology: A symposium honoring G. Ledyard Stebbins,* ed. L. Gottlieb and S. K. Jain, 275–303. London: Chapman and Hall.

Anttila, C. K., R. A. King, C. Ferris, D. R. Ayres, and D. R. Strong. 2000. Reciprocal hybrid formation of *Spartina* in San Francisco Bay. *Molecular Ecology* 9:765–770.

APHIS (Animal and Plant Health Inspection Service). 1997a. *USDA/APHIS permit 97 044 02r for field testing genetically engineered sugar beet plants. Environmental assessment and finding of no significant impact.*

APHIS (Animal and Plant Health Inspection Service). 1997b. *User's guide for introducing genetically engineered plants and microorganisms.* Technical bulletin 1783.

APHIS (Animal and Plant Health Inspection Service). 1998. *Appendix A. Determination of nonregulated status for transgenic glyphosate tolerant sugar beet line GTSB77.*

Arias, D. M., and L. H. Rieseberg. 1994. Gene flow between cultivated and wild sunflowers. *Theoretical and Applied Genetics* 89:655–660.

Arnold, M. L. 1993. *Iris nelsonii* (Iridaceae): Origin and genetic composition of a homoploid hybrid species. *American Journal of Botany* 80:577–583.

Arnold, M. L. 1997. *Natural hybridization and evolution.* New York: Oxford University Press.

Arnold, M. L., M. R. Bulger, J. M Burke, A. L. Hempel, and J. H. Williams. 1999. Natural hybridization: How low can you go and still be important? *Ecology* 80:371–381.

Arnold, M. L., and S. A. Hodges. 1995. Are natural hybrids fit or unfit relative to their parents? *Trends in Ecology and Evolution* 10:67–71.

Arriola, P. E., and N. C. Ellstrand. 1996. Crop-to-weed gene flow in the genus *Sorghum* (Poaceae): Spontaneous interspecific hybridization between johnsongrass, *Sorghum halepense,* and crop sorghum, *S. bicolor. American Journal of Botany* 83:1153–1160.

Arriola, P. E., and N. C. Ellstrand. 1997. Fitness of interspecific hybrids in the genus *Sorghum:* Persistence of crop genes in wild populations. *Ecological Applications* 7:512–518.

Ashburner, G. R., and H. C. Harries. 1999. Identifying markers for domestic-type coconut palms in segregating populations by applying generalised linear models to genetic resource data. In *Proceedings of the eleventh Australian plant breeding conference,* vol. 2, ed. P. Langridge, A. Barr, G. Auricht, G. Collins, A. Granger, D. Handford, and J. Paull, 77–78. Kent Town: Manning.

Ashworth, V. E. T. M., B. C. O'Brien, and E. A. Friar. 1999. Fingerprinting *Juniperus communis* L. cultivars using RAPD markers. *Madroño* 46:134–141.

Asker, S., and L. Jerling. 1992. *Apomixis in plants.* Boca Raton: CRC Press.

Astley, D., and J. G. Hawkes. 1979. The nature of the Bolivian weed potato species *Solanum sucrense. Euphytica* 28:685–696.

Austin, D. F. 1977. Hybrid polyploids in *Ipomoea* section *batatas. Journal of Heredity* 68:259–260.

Avise, J. C. 1994. *Molecular markers, natural history, and evolution.* New York: Chapman and Hall.

Baker, H. G. 1972. Migration of weeds. In *Taxonomy, phytogeography and evolution,* ed. D. H. Valentine, 327–347. London: Academic Press.

Baker, H. G. 1974. The evolution of weeds. *Annual Review of Ecology and Systematics* 5:1–23.

Ball, C. T., J. Keeley, H. Mooney, J. Seemann, and W. Winner. 1983. Relationship between form, function, and distribution of two *Arctostaphylos* species (Ericaceae) and their putative hybrids. *Acta Oecologia* 4:153–164.

Barham, W. S. 1953. The inheritance of a bitter principle in cucumbers. *Proceedings of the American Society for Horticultural Science* 62:441–442.

Barocka, K. H. 1985. Zucker- und Futterrüben. In *Lehrbuch der Pflanzenzüchtung landwirtschaftlicher Kulturformen, Bd. 2, Spezieller Teil,* ed. G. Fischbeck, W. Plarre, and W. Schuster, 245–287. Hamburg: Paul Parey.

Barr, A. R., and S. D. Tasker. 1992. Breeding herbicide resistance in oats: Opportunities and risks. In *Proceedings of the fourth international oat conference,* vol. 2, ed. A. R. Barr and R. W. Medd, 51–56. Adelaide: Fourth International Oat Conference, Inc.

Barrett, S. C. H. 1983. Crop mimicry in weeds. *Economic Botany* 37:255–282.

Barton, N. 2000. The rapid origin of reproductive isolation. *Science* 290:462–463.

Barton, N., and B. O. Bengtsson. 1986. The barrier to genetic exchange between hybridising populations. *Heredity* 56:357–376.

Barton, N. H., and G. M. Hewitt. 1985. Analysis of hybrid zones. *Annual Review of Ecology and Systematics* 16:113–148.

Bartsch, D., and U. Brand. 1998. Saline soil condition decreases rhizomania infection of *Beta vulgaris. Journal of Plant Pathology* 80:219–223.

Bartsch, D., U. Brand, C. Morak, M. Pohl-Orf, I. Schuphan, and N. C. Ellstrand. 2001. Biosafety of hybrids between sugar beet and Swiss chard. *Ecological Applications* 11:142–147.

Bartsch, D., and N. C. Ellstrand. 1999. Genetic evidence for the origin of Californian wild beets (genus *Beta*). *Theoretical and Applied Genetics* 99:1120–1130.

Bartsch, D., M. Lehnen, J. Clegg, M. Pohl-Orf, I. Schuphan, and N. C. Ellstrand. 1999. Impact of gene flow from cultivated beet on genetic diversity of wild sea beet populations. *Molecular Ecology* 8:1733–1741.

Bartsch, D., and M. Pohl-Orf. 1996. Ecological aspects of transgenic sugar beet: Transfer and expression of herbicide resistance in hybrids with wild beets. *Euphytica* 91:55–58.

Bartsch, D., and M. Schmidt. 1997. Influence of sugar beet breeding on populations of *Beta vulgaris* ssp. *maritima* in Italy. *Journal of Vegetation Science* 8:81–84.

Bates, D. M., Robinson, R. W., and C. Jeffrey. 1990. *The biology and utilization of the Cucurbitaceae*. Ithaca: Cornell University Press.

Baum, B. R. 1977. *Oats: Wild and cultivated*. Ottawa: Thorn Press.

Beebe, S., O. Toro, A. V. González, M. I. Chacón, and D. G. Debouck. 1997. Wild-weed-crop complexes of common bean (*Phaseolus vulgaris* L., Fabaceae) in the Andes of Peru and Colombia, and their implications for conservation and breeding. *Genetic Resources and Crop Evolution* 44:73–91.

Bell, R. L. 1990. Pears *(Pyrus)*. In *Genetic resources of temperate fruit and nut crops,* vol. 2, ed. J. N. Moore and J. R. Ballington, Jr., 655–699. Wageningen: International Society for Horticultural Science.

Bergelson, J. 1994. Changes in fecundity do not predict invasiveness: A model study of transgenic plants. *Ecology* 75:249–252.

Bergelson, J., and C. B. Purrington. 1996. Surveying patterns in the cost of resistance in plants. *American Naturalist* 148:536–558.

Bergelson, J., C. B. Purrington, and G. Wichmann. 1998. Promiscuity in transgenic plants. *Nature* 395:25.

Bierce, A. 1911. *The devil's dictionary*. Cleveland: World Publishing Co.

Bing, D. J., R. K. Downey, and G. F. W. Rakow. 1996. Hybridizations among *Brassica napus, B. rapa,* and *B. juncea* and their two weedy relatives *B. nigra* and *Sinapis arvensis* under open pollination conditions in the field. *Plant Breeding* 115:470–473.

Blancas, L. 2001. Hybridization between rare and common plant relatives: Implications for plant conservation genetics. Ph.D. diss., University of California, Riverside.

Blumler, M. A. 1998. Introgression of durum into wild emmer and the agricultural origin question. In *The origins of agriculture and crop domestication*, ed. A. B. Damania, J. Valkoun, G. Willcox, and C. O. Qualset, 252–268. Aleppo: ICARDA.

Bohac, J. B., P. D. Dukes, and D. F. Austin. 1995. Sweet potato. In *Evolution of crop plants*, ed. J. Smartt and N. W. Simmonds, 57–62. Harlow: Longman. 2nd ed.

Borrill, M. 1975. *Festuca*. In *Hybridization and the flora of the British Isles*, ed. C. A. Stace, 543–547. London: Academic Press.

Boudry, P., K. Broomberg, P. Saumitou-Laprade, M. Mörchen, J. Cuguen, and H. Van Dijk. 1994a. Gene escape in transgenic sugar beet: What can be learned from molecular studies of weed beet populations? *Proceedings of the third international symposium on the biosafety results of field tests of genetically modified plants and microorganisms*, ed. D. D. Jones, 75–87. Oakland: University of California, Division of Agriculture and Natural Resources.

Boudry, P., M. Mörchen, P. Saumitou-Laprade, P. Vernet, and H. Van Dijk. 1993. The origin and evolution of weed beets: Consequences for the breeding and release of herbicide-resistant transgenic sugar beets. *Theoretical and Applied Genetics* 87:471–478.

Boudry, P., R. Wieber, P. Saumitou-Laprade, K. Pillen, H. Van Dijk, and C. Jung. 1994b. Identification of RFLP markers closely linked to the bolting gene B and their significance for the study of the annual habit in beets (*Beta vulgaris* L.). *Theoretical and Applied Genetics* 88:852–858.

Broich, S. L., and R. G. Palmer. 1980. A cluster analysis of wild and domesticated soybean phenotypes. *Euphytica* 29:23–32.

Broich, S. L., and R. G. Palmer. 1981. Evolutionary studies of the soybean: The frequency and distribution of alleles among collections of *Glycine max* and *G. soja* of various origin. *Euphytica* 30:55–64.

Brubaker, C. L., J. A. Koontz, and J. F. Wendel. 1993. Bidirectional cytoplasmic and nuclear introgression in the New World cottons, *Gossypium barbadense* and *G. hirsutum* (Malvaceae). *American Journal of Botany* 80:1203–1208.

Brubaker, C. L., and J. F. Wendel. 1994. Reevaluating the origin of domesticated cotton (*Gossypium hirsutum*: Malvaceae) using nuclear restriction fragment length polymorphisms (RFLPs). *American Journal of Botany* 81:1309–1326.

Brunken, J., J. M. J. de Wet, and J. R. Harlan. 1977. The morphology and domestication of pearl millet. *Economic Botany* 31:163–174.

Buchmann, S. L., and G. P. Nabhan. 1996. *The forgotten pollinators*. Washington, D.C.: Island Press.

Buckler, E. S., and T. P. Holtsford. 1996. *Zea* systematics: Ribosomal ITS evidence. *Molecular Biology and Evolution* 13:612–622.

Burdon, J. J., D. R. Marshall, and J. D. Oates. 1992. Interactions between wild and cultivated oats in Australia. In *Proceedings of the fourth international oat conference*, vol. 2, ed. A. R. Barr and R. W. Medd, 82–87. Adelaide: Fourth International Oat Conference, Inc.

Burke, J. M., S. Tang, S. J. Knapp, and L. H. Rieseberg. 2002. Genetic analysis of sunflower domestication. *Genetics* 161:1257–1267.

Cade, T. J. 1983. Hybridization and gene exchange among birds in relation to conservation. In *Genetics and conservation: A reference for managing wild animal and plant populations*, ed. C. M. Schonewald-Cox, S. M. Chambers, B. MacBryde, and W. L. Thomas, 288–309. Menlo Park: Benjamin-Cummings.

Camazine, S., and R. A. Morse. 1988. The Africanized honeybee. *American Scientist* 76:465–471.

Carpenter, J. E. 2001. *Case studies in benefits and risks of agricultural biotechnology: Roundup Ready soybeans and Bt field corn*. Washington, D.C.: National Center for Food and Agricultural Policy.

Chang, T. T. 1995. Rice. In *Evolution of crop plants*, ed. J. Smartt and N. W. Simmonds, 147–155. Harlow: Longman. 2nd ed.

Charrier, A., and J. Berthaud. 1985. Botanical classification of coffee. In *Coffee: Botany, biochemistry, and production of beans and beverage*, ed. M. N. Clifford and K. C. Willson, 13–47. Westport: Avi Publishing.

Chazallon, G. 2000. Mesures complémentaires à prendre en vue de la production de semences tolérantes à un herbicide non sélectif. *Proceedings of the sixty-third IIRB congress*, 221–229.

Chèvre, A. M., F. Eber, A. Baranger, G. Hureau, P. Barret, H. Picault, and M. Renard. 1998. Characterization of backcross generation obtained under field conditions from oilseed rape–wild radish F_1 interspecific hybrids: An assessment of transgene dispersal. *Theoretical and Applied Genetics* 97:90–98.

Chu, Y. E., H. Morishima, and H. I. Oka. 1969. Reproductive barriers distributed in cultivated rice species and their wild relatives. *Japanese Journal of Genetics* 4:207–223.

Chu, Y. E., and H. I. Oka. 1970. Introgression across isolating barriers in wild and cultivated *Oryza* species. *Evolution* 24:344–355.

Cogolludo-Agustín, M., A. D. Agúndez, and L. Gil. 2000. Identification of native and hybrid elms in Spain using isozyme gene markers. *Heredity* 85:157–166.

Colwell, R. E., E. A. Norse, D. Pimentel, F. E. Sharples, and D. Simberloff. 1985. Genetic engineering in agriculture. *Science* 229:111–112.

Contant, R. B. 1976. Pyrethrum. In *Evolution of crop plants,* ed. N. W. Simmonds, 33–36. London: Longman.

Cook, R. J. 2000. Science-based risk assessment for the approval and use of plants in agricultural and other environments. In *Agricultural biotechnology and the poor,* ed. G. J. Persley and M. M. Lantin, 123–130. Washington, D.C.: Consultative Group on International Agricultural Research.

Crouch, M. L. 1998. *How the Terminator terminates: An explanation for the non-scientist of a remarkable patent for killing second generation seeds of crop plants.* Edmonds: Edmonds Institute.

Cuguen, J., R. Wattier, P. Saumitou-Laprade, D. Forcioli, M. Mörchen, H. Van Dijk, and P. Vernet. 1994. Gynodiocey and mitochondrial polymorphism in natural populations of the gynodioecious species *Beta vulgaris* ssp. *maritima. Genetics, Selection, and Evolution* 26:87–101.

Cummings, C. L., H. M. Alexander, and A. A. Snow. 1999. Increased pre-dispersal seed predation in sunflower crop-wild hybrids. *Oecologia* 121:330–338.

Dale, P. J., B. Clarke, and E. M. G. Fontes. 2002. Potential for the environmental impact of transgenic crops. *Nature Biotechnology* 20:567–574.

Daniell, H. 2002. Molecular strategies for gene containment in transgenic crops. *Nature Biotechnology* 20:581–586.

Daniell, H., R. Datta, S. Varma, S. Gray, and S.-B. Lee. 1998. Containment of herbicide resistance through genetic engineering of chloroplast genome. *Nature Biotechnology* 16:345–348.

Daniels, J., and B. T. Roach. 1987. Taxonomy and evolution. In *Sugarcane improvement through breeding,* ed. D. J. Heinz, 7–83. Amsterdam: Elsevier.

Darmency, H., E. Lefol, and A. Fleury. 1998. Spontaneous hybridizations between oilseed rapes and wild radish. *Molecular Ecology* 7:1467–1473.

Dave, B. B. 1943. The wild rice problem in the Central Provinces and its solution. *Indian Journal of Agriculture Sciences* 13:46–53.

Davies, D. R. 1995. Peas. In *Evolution of crop plants,* ed. J. Smartt and N. W. Simmonds, 294–296. Harlow: Longman. 2nd ed.

Day, P. R. 1987. Concluding remarks by Dr. Peter R. Day. In *Regulatory considerations: Genetically-engineered plants.* San Francisco: Center for Science Information.

Debouck, D. G., and J. Smartt. 1995. Beans. In *Evolution of crop plants,* ed. J. Smartt and N. W. Simmonds, 287–294. Harlow: Longman. 2nd ed.

Debouck, D. G., O. Toro, O. M. Paredes, W. C. Johnson, and P. Gepts. 1993. Genetic diversity and ecological distribution of *Phaseolus vulgaris* in northwestern South America. *Economic Botany* 47:408–423.

De Candolle, A. 1886. *Origin of cultivated plants*. 2nd ed. New York: Hafner (reprinted 1959).

DeJoode, D. R., and J. F. Wendel. 1992. Genetic diversity and origin of the Hawaiian Islands cotton, *Gossypium tomentosum*. *American Journal of Botany* 79:1311–1319.

Delgado Salinas, A., A. Bonet, and P. Gepts. 1988. The wild relative of *Phaseolus vulgaris* in Middle America. In *Genetic resources of* Phaseolus *beans*, ed. P. Gepts, 163–184. Dordrecht: Kluwer.

Demeke, T., and R. P. Adams. 1994. The use of PCR-RAPD analysis in plant taxonomy and evolution. In *PCR technology: Current innovations*, ed. H. G. Griffin and A. M. Griffin, 179–191. Boca Raton: CRC Press.

Den Nijs, T. P. M., and S. J. Peloquin. 1977. 2n gametes in potato species and their function in sexual polyploidization. *Euphytica* 26:585–600.

Derick, R. A. 1933. Natural crossing with wild oats, *Avena fatua*. *Scientific Agriculture* 13:458–459.

Desplanque, B., P. Boudry, K. Broomberg, P. Saumitou-Laprade, J. Cuguen, and H. Van Dijk. 1999. Genetic diversity and gene flow between wild, cultivated and weedy forms of *Beta vulgaris* L. (Chenopodiaceae), assessed by RFLP and microsatellite markers. *Theoretical and Applied Genetics* 98:1194–1201.

DeVries, F. T., R. Van der Meijden, and W. A. Brandenburg. 1992. Botanical files: A study of the real chances for spontaneous gene flow from cultivated plants to the wild flora of the Netherlands. *Gorteria,* suppl. 1:1–100.

De Wet, J. M. J. 1995a. Finger millet. In *Evolution of crop plants*, ed. J. Smartt and N. W. Simmonds, 137–140. Harlow: Longman. 2nd ed.

De Wet, J. M. J. 1995b. Pearl millet. In *Evolution of crop plants*, ed. J. Smartt and N. W. Simmonds, 156–159. Harlow: Longman. 2nd ed.

De Wet, J. M. J., and J. R. Harlan. 1975. Weeds and domesticates: Evolution in the man-made habitat. *Economic Botany* 29:99–107.

Diaz, J., P. Schmiediche, and D. F. Austin. 1996. Polygon of crossability between eleven species of *Ipomoea*: Section Batatas. *Euphytica* 88:189–200.

Dickson, E. E., S. Kresovich, and N. F. Weeden. 1991. Isozymes in North American *Malus* (Rosaceae): Hybridization and species differentiation. *Systematic Botany* 16:363–375.

Dietz-Pfeilstetter, A., and M. Kirchner. 1998. Analysis of gene inheritance and expression in hybrids between transgenic sugar beet and wild beets. *Molecular Ecology* 7:1693–1700.

DiFazio, S. P., S. Leonardi, S. Cheng, and S. H. Strauss. 1999. Assessing potential risks of transgene escape from fiber plantations. In *Gene flow and agriculture—Relevance for transgenic crops*, ed. P. J. W. Lutman, 171–176. Staffordshire: British Crop Protection Council.

Dobzhansky, T. 1953. Natural hybrids of two species of *Arctostaphylos* in the Yosemite region of California. *Heredity* 7:73–79.

Doebley, J. 1989. Isozymic evidence and the evolution of crop plants. In *Isozymes in plant biology*, ed. D. E. Soltis and P. S. Soltis, 165–191. Portland: Dioscorides.

Doebley, J. 1990. Molecular evidence for gene flow among *Zea* species. *BioScience* 40:443–448.

Doebley, J. 1996. Genetic dissection of the morphological evolution of maize. *Aliso* 14:297–304.

Doggett, H., and B. N. Majisu. 1968. Disruptive selection in crop development. *Heredity* 23:1–23.

Doggett, H., and K. E. Prasada Rao. 1995. Sorghum. In *Evolution of crop plants,* ed. J. Smartt and N. W. Simmonds, 173–180. Harlow: Longman. 2nd ed.

Dorado, O., L. H. Rieseberg, and D. M. Arias. 1992. Chloroplast DNA introgression in southern California sunflowers. *Evolution* 46:566–572.

Dow, B. D., and M. V. Ashley. 1998. High levels of gene flow in bur oak revealed by paternity analysis using microsatellites. *Journal of Heredity* 89:62–70.

Duckett, J. G., and C. N. Page. 1975. *Equisetum* L. In *Hybridization and the flora of the British Isles,* ed. C. A. Stace, 99–103. London: Academic Press.

Durant, A. 1976. Flax and linseed. In *Evolution of crop plants,* ed. N. W. Simmonds, 190–193. London: Longman.

Duvick, D. N. 1984. Genetic diversity in major farm crops on the farm and in reserve. *Economic Botany* 38:161–178.

Ellstrand, N. C. 1988. Pollen as a vehicle for the escape of engineered genes? In *Planned release of genetically engineered organisms,* ed. J. Hodgson and A. M. Sugden, S30–S32. Cambridge: Elsevier.

Ellstrand, N. C. 1989. Gene rustlers. *Omni* 11(7): 33.

Ellstrand, N. C. 1992. Gene flow among seed plant populations. *New Forests* 6:241–256.

Ellstrand, N. C. 2000. The elephant that is biotechnology: Comments on "Genetically Modified Crops: Risks and Promise" by Gordon Conway. *Conservation Ecology* 4(1): 8. http://www.consecol.org/vol4/iss1/art8.

Ellstrand, N. C. 2001. When transgenes wander, should we worry? *Plant Physiology* 125:1543–1545.

Ellstrand, N. C., B. Devlin, and D. L. Marshall. 1989. Gene flow by pollen into small populations: Data from experimental and natural stands of wild radish. *Proceedings of the National Academy of Sciences* 86:9044–9047.

Ellstrand, N. C., and D. R. Elam. 1993. Population genetic consequences of small population size: Implications for plant conservation. *Annual Review of Ecology and Systematics* 24:217–242.

Ellstrand, N. C., and C. A. Hoffman. 1990. Hybridization as an avenue for escape of engineered genes. *BioScience* 40:438–442.

Ellstrand, N. C., J. M. Lee, B. O. Bergh, M. D. Coffey, and G. A. Zentmyer. 1986. Isozymes confirm hybrid parentage for "G755" selections. *California Avocado Society Yearbook* 70:199–203.

Ellstrand, N. C., J. M. Lee, J. E. Keeley, and S. C. Keeley. 1987. Ecological isolation and introgression: Biochemical confirmation of introgression in an *Arctostaphylos* (Ericaceae) population. *Acta Oecologia/Oecologia Plantarum* 8:299–308.

Ellstrand, N. C., H. C. Prentice, and J. F. Hancock. 1999. Gene flow and introgression from domesticated plants into their wild relatives. *Annual Review of Ecology and Systematics* 30:539–563.

Ellstrand, N. C., H. C. Prentice, and J. F. Hancock. 2002. Gene flow and introgression from domesticated plants into their wild relatives. In *Horizontal gene transfer,* ed. M. Syvanen and C. I. Kado, 217–234. San Diego: Academic Press. 2nd ed.

Ellstrand, N. C., and K. Schierenbeck. 2000. Hybridization as a stimulus for the evolution of invasiveness in plants? *Proceedings of the National Academy of Sciences* 97:7043–7050.

Ellstrand, N. C., R. Whitkus, and L. H. Rieseberg. 1996. Distribution of spontaneous plant hybrids. *Proceedings of the National Academy of Sciences* 93:5090–5093.

Epling, C. 1947. The genetic aspects of natural populations—Actual and potential gene flow in natural populations. *American Naturalist* 81:104–113.

Ervin, D. E., S. E. Batie, R. Welsh, C. L. Carpentier, J. I. Fern, N. J. Richman, and M. A. Schulz. 2000. *Transgenic crops: An environmental assessment.* Winrock: Henry A. Wallace Center for Agricultural and Environmental Policy.

Evans, G. E. 1995. Rye. In *Evolution of crop plants,* ed. J. Smartt and N. W. Simmonds, 166–170. Harlow: Longman. 2nd ed.

Fantes, J. A., and G. R. Mackay. 1978. The production of disomic addition lines of *Brassica campestris. Cruciferae Newsletter* 4:36–37.

Fehr, W. R. 1987. *Principles of cultivar development.* Vol. 1, *Theory and technique.* New York: Macmillan.

Feldman, M., F. G. H. Lupton, and T. E. Miller. 1995. Wheats. In *Evolution of crop plants,* ed. J. Smartt and N. W. Simmonds, 184–192. Harlow: Longman. 2nd ed.

Fenster, C. B., and L. F. Galloway. 2000. Inbreeding and outbreeding depression in natural populations of *Chamaecrista fasciculata. Conservation Biology* 14:1406–1412.

Fisher, R. A. 1937. On the wave of advance of an advantageous allele. *Annals of Eugenics* 7:355–369.

Flake, R. H., E. von Rudloff, and B. L. Turner. 1969. Quantitative study of clinal variation in *Juniperus virginiana* using terpenoid data. *Proceedings of the National Academy of Sciences, USA* 64:487–494.

Forcioli, D., P. Saumitou-Laprade, G. Michaelis, and J. Cuguen. 1994. Chloroplast DNA polymorphism revealed by a fast, non-radioactive method in *Beta vulgaris* subsp. *maritima. Molecular Ecology* 3:173–176.

Ford-Lloyd, B. 1995. Sugarbeet, and other cultivated beets. In *Evolution of crop plants,* ed. J. Smartt and N. W. Simmonds, 35–40. Harlow: Longman. 2nd ed.

Frank, S. A., and M. Slatkin. 1992. Fisher's fundamental theorem of natural selection. *Trends in Ecology and Evolution* 7:92–95.

Frankel, R., and E. Galun. 1977. *Pollination mechanisms, reproduction and plant breeding.* Berlin: Springer-Verlag.

Franz, J. E., M. K. Mao, and J. A. Sikorski. 1997. *Glyphosate: A unique global herbicide.* Washington, D.C.: American Chemical Society.

Freyre, R., R. Ríos, L. Guzmán, D. G. DeBouck, and P. Gepts. 1996. Ecogeographic distribution of *Phaseolus* ssp. (Fabaceae) in Bolivia. *Economic Botany* 50:195–215.

Friedman, S. T., and W. T. Adams. 1985. Estimation of gene flow into two seed orchards of loblolly pine (*Pinus taeda* L.). *Theoretical and Applied Genetics* 69:609–615.

Fryxell, P. A. 1957. Mode of reproduction of higher plants. *Botanical Review* 23:135–233.

Fryxell, P. A. 1979. *The natural history of the cotton tribe (Malvaceae, tribe Gossypieae).* College Station: Texas A&M University Press.

Futuyma, D. 1998. *Evolutionary biology.* Sunderland: Sinauer. 3rd ed.

Gavrilets, S., and M. B. Cruzan. 1998. Neutral gene flow across single locus clines. *Evolution* 52:1277–1284.

Gepts, P. 2001. Origins of plant agriculture and major crop plants. In *Our fragile world: Challenges and opportunities for sustainable development,* ed. M. K. Tolba, 629–637. Oxford: EOLSS Publishers.

Gill, K. S. 1989. Germplasm collections and the public plant breeder. In *The use of plant genetic resources,* ed. A. H. D. Brown, O. H. Frankel, D. R. Marshall, and J. T. Williams, 3–16. Cambridge: Cambridge University Press.

Goodman, M. M. 1995. Maize. In *Evolution of crop plants,* ed. J. Smartt and N. W. Simmonds, 192–202. Harlow: Longman. 2nd ed.

Goodman, R. M., and N. Newell. 1985. Genetic engineering of plants for herbicide resistance: Status and prospects. In *Engineered organisms the environment: Scientific issues,* ed. H. O. Halvorson, D. Pramer, and M. Rogul, 47–53. Washington, D.C.: American Society for Microbiology.

Gottlieb, L. D. 1972. Levels of confidence in the analysis of hybridization. *Annals of the Missouri Botanical Garden* 59:435–446.

Gottlieb, L. D. 1982. Conservation and duplication of isozymes in plants. *Science* 216:373–380.

Grant, V. 1964a. The biological composition of a taxonomic species in *Gilia. Advances in Genetics* 12:281–328.

Grant, V. 1964b. Genetic and taxonomic studies in *Gilia.* XII. Fertility relationships of the polyploid cobwebby gilias. *Aliso* 5:479–507.

Grant, V. 1975. *Genetics of flowering plants.* New York: Columbia University Press.

Grant, V. 1981. *Plant speciation.* New York: Columbia University Press. 2nd ed.

Grant, V., and A. Grant. 1960. Genetic and taxonomic studies in *Gilia.* XI. Fertility relationships of the diploid cobwebby gilias. *Aliso* 4:435–481.

Gray, A. J. 1986. Do invading species have definable genetic characteristics? *Philosophical Transactions of the Royal Society of London* B 314:655–674.

Gray, A. J., D. F. Marshall, and A. F. Raybould. 1991. A century of evolution in *Spartina anglica. Advances in Ecological Research* 21:1–61.

Great Plains Flora Association. 1986. *Flora of the Great Plains.* Lawrence: University of Kansas Press.

Gressel, J. 1991. Why get herbicide resistance? It can be prevented or delayed. In *Herbicide resistance in weeds and crops,* ed. J. C. Caseley, G. W. Cussans, and R. K. Atkin, 1–26. Oxford: Butterworth-Heinemann.

Gressel, J. 1999. Tandem constructs: Preventing the rise of superweeds. *Trends in Biotechnology* 17:361–366.

Grumet, R. 1995. Genetic engineering for crop virus resistance. *HortScience* 30:449–456.

Guadagnuolo, R., D. Savova-Bianchi, and F. Felber. 2001a. Gene flow from wheat (*Triticum aestivum* L.) to jointed goatgrass (*Aegilops cylindrica* Host.), as revealed by RAPD and microsatellite markers. *Theoretical and Applied Genetics* 103:1–8.

Guadagnuolo, R., D. Savova-Bianchi, J. Keller-Senften, and F. Felber. 2001b. Search for evidence of introgression of wheat (*Triticum aestivum* L.) traits into sea barley (*Horedum marinum* s. str. Huds.) and bearded wheatgrass (*Elymus caninus* L.) in central and northern Europe, using isozymes, RAPD, and microsatellite markers. *Theoretical and Applied Genetics* 103:191–196.

Hadidi, A., R. K. Khetarpal, and H. Koganezawa. 1998. *Plant virus disease control.* St. Paul: APS Press.

Hails, R. S. 2000. Genetically modified plants—The debate continues. *Trends in Ecology and Evolution* 15:14–18.

Halfhill, M. D., H. A. Richards, S. A. Mahon, and C. N. Stewart, Jr. 2001. Expression of GFP and Bt transgenes in *Brassica napus* and hybridization with *Brassica rapa. Theoretical and Applied Genetics* 103:659–667.

Hall, L., K. Topinka, J. Huffman, L. Davis, and A. Allen. 2000. Pollen flow between herbi-cide-resistant *Brassica napus* is the cause of multiple-resistant *B. napus* volunteers. *Weed Science* 48:688–694.

Hall, M. T. 1952. Variation and hybridization in *Juniperus*. *Annals of the Missouri Botani-cal Garden* 39:1–64.

Hancock, J. F. 1992. *Plant evolution and the origin of crop species*. Englewood Cliffs: Prentice-Hall.

Hancock, J. F., R. Grumet, and S. C. Hokanson. 1996. The opportunity for escape of en-gineered genes from transgenic crops. *HortScience* 31:1080–1085.

Handel, S. N. 1983. Pollination ecology, plant population structure, and gene flow. In *Pollination biology*, ed. L. Real, 163–212. New York: Academic Press.

Hanneman, R. E. 1994. The testing and release of transgenic potatoes in the North American center of diversity. In *Biosafety for sustainable agriculture: Sharing biotechnol-ogy regulatory experiences of the Western Hemisphere*, ed. A. F. Krattiger and A. Rosemarin, 47–67. Ithaca: ISAAA and Stockholm: SEI.

Hansen, L. B., H. R. Siegismund, and R. B. Jørgensen. 2001. Introgression between oil-seed rape (*Brassica napus* L.) and its weedy relative *B. rapa* in a natural population. *Genetic Resources and Crop Evolution* 48:621–627.

Hanson, M. 1993. Dispersed unidirectional introgression from *Yucca schidigera* into *Yucca baccata* (Agavaceae). Ph.D. diss., Claremont Graduate School, Claremont, Ca-lif.

Harberd, D. J. 1975. *Brassica* L. In *Hybridization and the flora of the British Isles*, ed. C. A. Stace, 137–139. London: Academic Press.

Hardon, J. J. 1995. Oil palm. In *Evolution of crop plants,* ed. J. Smartt and N. W. Simmonds, 395–399. Harlow: Longman. 2nd ed.

Hardon, J. J., and G. Y. Tan. 1969. Interspecific hybrids in the genus *Elaeis*. I. Crossibility, cytogenetics, and fertility of the F_1 hybrids. *Euphytica* 18:380–388.

Hare, P. D., and N.-H. Chua. 2002. Excision of selectable marker genes from transgenic plants. *Nature Biotechnology* 20:575–580.

Harlan, J. R. 1965. The possible role of weedy races in the evolution of cultivated plants. *Euphytica* 14:173–176.

Harlan, J. R. 1995. Barley. In *Evolution of crop plants*, ed. J. Smartt and N. W. Simmonds, 140–147. Harlow: Longman. 2nd ed.

Harlan, J. R., J. M. J. de Wet, and E. G. Price. 1973. Comparative evolution of cereals. *Evolution* 27:311–325.

Harley, R. M. 1975. *Mentha* L. In *Hybridization and the flora of the British Isles,* ed. C. A. Stace, 383–390. London: Academic Press.

Harries, H. C. 1995. Coconut. In *Evolution of crop plants,* ed. J. Smartt and N. W. Simmonds, 57–62. Harlow: Longman. 2nd ed.

Harrison, R. G. 1990. Hybrid zones: Windows on evolutionary process. *Oxford Surveys in Evolutionary Biology* 7:69–128.

Hartl, D. L., and A. G. Clark. 1997. *Principles of population genetics*. Sunderland: Sinauer. 3rd ed.

Haslberger, A. 2001. GMO contamination of seeds. *Nature Biotechnology* 19:613.

Hauser, T. P., R. B. Jørgensen, and H. Østergård. 1998a. Fitness of backcross and F_2 hy-brids between weedy *Brassica rapa* and oilseed rape *(B. napus)*. *Heredity* 81:436–443.

Hauser, T. P., R. G. Shaw, and H. Østergård. 1998b. Fitness of F_1 hybrids between weedy *Brassica rapa* and oilseed rape *(B. napus)*. *Heredity* 81:429–435.

Hawkes, J. G. 1990. *The potato: Evolution, biodiversity, and genetic resources.* London: Bellhaven Press.

Hawkes, J. G., and J. P. Hjerting. 1989. *The potatoes of Bolivia: Their breeding value and evolutionary relationships.* Oxford: Oxford University Press.

Hedrick, P. W. 2000. *Genetics of populations.* Sudbury: Jones and Bartlett.

Hegde, S. G., and J. G. Waines. 2002. Hybridization and introgression between bread wheat (*Triticum aestivum* L.) and wild and weedy relatives. *Crop Science.* In press.

Heiser, C. B. 1978. Taxonomy of *Helianthus* and the origin of domesticated sunflower. In *Sunflower Science and Technology,* ed. J. F. Carter, 31–54. Madison: American Society of Agronomy, Crop Science Society, and Soil Science Society of America.

Hickman, J. C. 1993. *The Jepson manual: Higher plants of California.* Berkeley: University of California Press.

Hilu, K. 1983. The role of single-gene mutation in the evolution of flowering plants. *Evolutionary Biology* 16:97–128.

Holm, L., J. Doll, E. Holm, J. Pancho, and J. Herberger. 1997. *World weeds: Natural histories and distribution.* New York: John Wiley and Sons.

Holm, L. G., D. L. Plucknett, J. V. Pancho, and J. P. Herberger. 1977. *The world's worst weeds: Distribution and biology.* Honolulu: University Press of Hawaii.

Hood, E. E., D. R. Witcher, S. Maddock, T. Meyer, C. Baszczynski, M. Bailey, P. Flynn, J. Register, L. Marshall, D. Bond, E. Kulisek, A. Kusnadi, R. Evangelista, Z. Nikolov, C. Wooge, R. J. Mehigh, R. Herman, W. K. Kappel, D. Ritland, C. P. Li, and J. A. Howard. 1997. Commercial production of avidin from transgenic maize: Characterization of transformant, production, processing, extraction, and purification. *Molecular Breeding* 3:291–306.

Hopper, S. D. 1995. Evolutionary networks: Natural hybridization and its conservation significance. In *Nature conservation 4: The role of networks,* ed. D. A. Saunders, J. L. Craig, and E. M. Mattiske, 51–66. Chipping Norton: Surrey Beatty and Sons.

Hussendoerfer, E. 1999. Short note: Identification of natural hybrids *Juglans* × *intermedia* CARR. using isoenzyme gene markers. *Silvae Genetica* 48:50–52.

Huxel, G. R. 1999. Rapid displacement of native species by invasive species: Effect of hybridization. *Biological Conservation* 89:143–152.

Jain, S. K. 1977. Genetic diversity of weedy rye populations in California. *Crop Science* 17:480–482.

Jarvis, D. J., and T. Hodgkin. 1999. Wild relatives and crop cultivars: Detecting natural introgression and farmer selection of new genetic combinations in agroecosystems. *Molecular Ecology* 8:S159–S173.

Jenczewski, E., J.-M. Prosperi, and J. Ronfort. 1999. Evidence for gene flow between wild and cultivated *Medicago sativa* (Leguminosae) based on allozyme markers and quantitative traits. *American Journal of Botany* 86:677–687.

Jennings, D. L. 1995. Cassava. In *Evolution of crop plants,* ed. J. Smartt and N. W. Simmonds, 128–132. Harlow: Longman. 2nd ed.

Jepson, W. L. 1909. *A flora of California.* San Francisco: Cunningham, Curtiss, and Welch.

Jiang, J., B. Freibe, and B. S. Gill. 1994. Recent advances in alien gene transfer in wheat. *Euphytica* 73:199–212.

Johns, T., Z. Huamán,, C. Ochoa, and P. E. Schiediche. 1987. Relationships among wild, weed, and cultivated potatoes in the *Solanum* × *ajanhuiri* complex. *Systematic Botany* 40:409–552.

Johnson, R. T., and L. M. Burtch. 1958. The problem of wild annual sugar beets in California. *Journal of the American Society of Sugar Beet Technologists* 10:311–317.

Jørgensen, R. B., and B. Andersen. 1994. Spontaneous hybridization between oilseed rape *(Brassica napus)* and weedy *Brassica campestris* (Brassicaceae): A risk of growing genetically modified oilseed rape. *American Journal of Botany* 81:1620–1626.

Jørgensen, R. B., B. Andersen, T. P. Hauser, L. Landbo, T. R. Mikkelsen, and H. Østergård. 1998. Introgression of crop genes from oilseed rape *(Brassica napus)* to related wild species—An avenue for the escape of engineered genes. In *Proceedings of the international symposium on Brassicas,* ed. T. Grégoire and A. A. Monteiro, 211–217. Rennes: ISHS.

Jørgensen, R. B., B. Andersen, L. Landbo, and T. Mikkelsen. 1996. Spontaneous hybridization between oilseed rape *(Brassica napus)* and weedy relatives. In *Proceedings of the international symposium on Brassicas/Ninth crucifer genetics workshop,* ed. J. S. Dias, I. Crute, and A. A. Monteiro, 193–197. Lisbon: ISHS.

Kareiva, P., and M. Marvier. 2000. An overview of risk assessment procedures applied to genetically engineered crops. In *Incorporating science, economics, and sociology in developing sanitary and phytosanitary standards in international trade,* 231–238. Washington, D.C.: National Academy Press.

Kareiva, P., I. M. Parker, and M. Pascual. 1996. Can we use experiments and models in predicting the invasiveness of genetically engineered organisms? *Ecology* 77:1670–1675.

Kato, T. A. 1997. Review of introgression between maize and teosinte. In *Gene flow among maize landraces, improved maize varieties, and teosinte: Implications for transgenic maize,* ed. J. A. Serratos, M. C. Willcox, and F. Castillo González, 44–53. Mexico, DF: CIMMYT.

Keeler, K. H., and C. E. Turner. 1990. Management of transgenic plants in the environment. In *Risk assessment in genetic engineering,* ed. M. Levin and H. Strauss, 189–218. New York: McGraw-Hill.

Keenan, R. J., and W. P. C. Stemmer. 2002. Nontransgenic crops from transgenic plants. *Nature Biotechnology* 20:215–216.

Keim, P., R. C. Shoemaker, and R. G. Palmer. 1989. Restriction fragment length polymorphism diversity in soybean. *Theoretical and Applied Genetics* 77:786–792.

Kelly, A. F., and R. A. T. George. 1998. *Encyclopaedia of seed production of world crops.* Chichester: John Wiley and Sons.

Kermicle, J. 1997. Cross compatibility within the genus *Zea*. In *Gene flow among maize landraces, improved maize varieties, and teosinte: Implications for transgenic maize,* ed. J. A. Serratos, M. C. Willcox, and F. Castillo González, 40–43. Mexico, DF: CIMMYT.

Kernick, M. D. 1961. Seed production of specific crops. In *Agricultural and horticultural seeds,* 181–461. Rome: Food and Agriculture Organization of the United Nations.

Kiang, Y. T., J. Antonovics, and L. Wu. 1979. The extinction of wild rice *(Oryza perennis formosana)* in Taiwan. *Journal of Asian Ecology* 1:1–9.

Kilman, S. 2001. Food industry shuns bioengineered sugar. *Wall Street Journal,* 27 April, p. B5.

Kimber, G., and M. Feldman. 1987. *Wild wheat.* Special report 353. Columbia: University of Missouri.

Kirkpatrick, K. J., and H. D. Wilson. 1988. Interspecific gene flow in Cucurbita: *C. texana* vs. *C. pepo. American Journal of Botany* 75:519–527.

Klinger, T. 2002. Variability and uncertainty in crop-to-wild hybridization. In *Genetically engineered organisms: Assessing environmental and human health effects*, ed. D. K. Letourneau and B. E. Burrows, 1–15. Boca Raton: CRC Press.

Klinger, T., P. E. Arriola, and N. C. Ellstrand. 1992. Crop-weed hybridization in radish (*Raphanus sativus* L.): Effects of distance and population size. *American Journal of Botany* 79:1431–1435.

Klinger, T., D. R. Elam, and N. C. Ellstrand. 1991. Radish as a model system for the study of engineered gene escape rates via crop-weed mating. *Conservation Biology* 5:531–535.

Klinger, T., and N. C. Ellstrand. 1994. Engineered genes in wild populations: Fitness of weed-crop hybrids of radish, *Raphanus sativus* L. *Ecological Applications* 4:117–120.

Klinger, T., and N. C. Ellstrand. 1999. Transgene movement via gene flow: Recommendations for improved biosafety assessment. In *Methods for risk assessment of transgenic plants. Vol. 3, Ecological risks and prospects of transgenic plants, where do we go from here? A dialogue between biotech industry and science,* ed. K. Ammann, Y. Jacot, G. Simonsen, and G. Kjellsson, 129–141. Basel: Birkhauser.

Knowles, P. F., and A. Ashri. 1995. Safflower. In *Evolution of crop plants,* ed. J. Smartt and N. W. Simmonds, 47–50. Harlow: Longman. 2nd ed.

Kramer, K. J., T. D. Morgan, J. E. Throne, F. E. Dowell, M. Bailey, and J. A. Howard. 2000. Transgenic maize expressing avidin is resistant to storage insect pests. *Nature Biotechnology* 18:670–674.

Kramer, M. G., and K. Redenbaugh. 1994. Commercialization of a tomato with an antisense polygalacturonase gene: The FLAVR SAVR–TM tomato story. *Euphytica* 79:293–297.

Kuehn, M., M. J. E. Marcinko, and B. N. White. 1999. An examination of hybridization between the cattail species *Typha latifolia* and *Typha angustifolia* using random amplified polymorphic DNA and chloroplast DNA markers. *Molecular Ecology* 8:1981–1990.

Kuvshinov, V., K. Koivu, A. Kanerva, and E. Pehu. 2001. Molecular control of transgene escape from genetically modified plants. *Plant Science* 160:517–522.

Ladizinsky, G. 1985. Founder effect in crop-plant evolution. *Economic Botany* 39:191–199.

Ladizinsky, G. 1989. Ecological and genetic considerations in collecting and using wild relatives. In *The use of plant genetic resources,* ed. A. H. D. Brown, O. H. Frankel, D. R. Marshall, and J. T. Williams, 297–305. Cambridge: Cambridge University Press.

Ladizinsky, G. 1992. Crossibility relations. In *Distant hybridization of crop plants,* ed. G Kalloo and J. B. Chowdhury, 15–31. Berlin: Springer-Verlag.

Ladizinsky, G. 1995. Chickpea. In *Evolution of crop plants,* ed. J. Smartt and N. W. Simmonds, 258–261. Harlow: Longman. 2nd ed.

Ladizinsky, G. 1998. *Plant evolution under domestication.* Dordrecht: Kluwer.

Ladizinsky, G., and A. Adler. 1976. Genetic relationships among the annual species of *Cicer* L. *Theoretical and Applied Genetics* 48:197–203.

Lande, R. 1988. Genetics and demography in biological conservation. *Science* 241:1455–1459.

Landrum, L. R., W. D. Clark, W. P. Sharp, and J. Brendecke. 1995. Hybridization between *Psidium guajava* and *P. guineense* (Myrtaceae). *Economic Botany* 49:153–161.

Langevin, S., K. Clay, and J. B. Grace. 1990. The incidence and effects of hybridization between cultivated rice and its related weed red rice (*Oryza sativa* L.) *Evolution* 44:1000–1008.

Ledig, F. T. 1992. Human impacts on genetic diversity in forest ecosystems. *Oikos* 63:87–108.

Lefèvre, F., and A. Charrier. 1993. Isozyme diversity within African *Manihot* germplasm. *Euphytica* 6673–6680.

Lefol, E., V. Danielou, and H. Darmency. 1996a. Predicting hybridization between transgenic oilseed rape and mustard. *Field Crops Research* 45:153–161.

Lefol, E., A. Fleury, and H. Darmency. 1996b. Gene dispersal from transgenic crops. II. Hybridization between oilseed rape and the wild hoary mustard. *Sexual Plant Reproduction* 9:189–196.

Lenormand, T. 2002. Gene flow and the limits to natural selection. *Trends in Ecology and Evolution* 17:183–189.

Letschert, J. P. W. 1993. *Beta* section *Beta,* biogeographical patterns of variation and taxonomy. Ph.D. diss. Wageningen Agricultural University Papers 93-1. Wageningen: University of Wageningen.

Levin, D. A. 1978. The origin of isolating mechanisms in flowering plants. *Evolutionary Biology* 11:185–317.

Levin, D. A. 1981. Dispersal versus gene flow in plants. *Annals of the Missouri Botanical Garden* 68:233–253.

Levin, D. A. 1984. Immigration in plants: An exercise in the subjunctive. In *Perspectives on plant population ecology,* ed. R. Dirzo and J. Sarukhán, 242–260. Sunderland: Sinauer.

Levin, D. A. 2001. The congener as an agent of extermination and rescue of rare species. In *Evolutionary conservation biology,* ed. R. Ferriere, U. Dieckmann, and D. Couvet. Laxenburg: International Institute for Applied Systems Analysis. In press.

Levin, D. A., J. Francisco-Ortega, and R. K. Jansen. 1996. Hybridization and the extinction of rare plant species. *Conservation Biology* 10:10–16.

Levin, D. A., and H. W. Kerster. 1974. Gene flow in seed plants. *Evolutionary Biology* 7:139–220.

Lev-Yadun, S., and R. Sederoff. 2001. Grafting for transgene containment. *Nature Biotechnology* 19:1104.

Lewis, E. J. 1975. *Festuca* L. × *Lolium* L. = × *Festulolium* Aschers & Graebn. In *Hybridization and the flora of the British Isles,* ed. C. A. Stace, 547–552. London: Academic Press.

Lewontin, R. C., and L. C. Birch. 1966. Hybridization as a source of variation for adaptation to new environments. *Evolution* 20:315–336.

Li, S.-G., Y. L. Liu, F. Zhu, Y.-Y. Luo, L.-Y. Kang, and B. Tian. 1997. Genetically engineered male sterile tobacco plants and their sensitivity to temperature. *Acta Botanica Sinica* 39:231–235.

Linder, C. R. 1998. Potential persistence of transgenes: Seed performance of transgenic canola and wild × canola hybrids. *Ecological Applications* 8:1180–1195.

Linder, C. R., I. Taha, G. J. Seiler, A. A. Snow, and L. H. Rieseberg. 1998. Long-term introgression of crop genes into wild sunflower populations. *Theoretical and Applied Genetics* 96:339–347.

Linhart, Y. B., and M. C. Grant. 1996. Evolutionary significance of local genetic differentiation in plants. *Annual Review of Ecology and Systematics* 27:237–277.

Longden, P. C. 1989. Effect of increasing weed-beet density on sugar-beet yield and quality. *Annals of Applied Biology* 114:527–532.

Longden, P. C. 1993. Weed beet: A review. *Aspects of Applied Biology* 35:185–194.

Lönn, M., H. C. Prentice, and K. Bengtsson. 1996. Genetic structure, allozyme-habitat associations, and reproductive fitness in *Gypsophila fastigiata* (Caryophyllaceae). *Oecologia* 106:308–316.

Lord, E. M. 1981. Cleistogamy: A tool for plant morphogenesis, function, and evolution. *Botanical Review* 47:421–429.

Love, S. 1994. Ecological risk of growing transgenic potatoes in the United States and Canada. *American Potato Journal* 71:647–658.

Lowe, A. J., C. M. Gillies, J. Wilson, and I. K. Dawson. 2000. Conservation genetics of bush mango from Central/West Africa: Implications from random amplified polymorphic DNA analysis. *Molecular Ecology* 9:831–841.

Luby, J. J., and R. J. McNichol. 1995. Gene flow from cultivated to wild raspberries in Scotland: Developing a basis for risk assessment for testing and deployment of transgenic cultivars. *Theoretical and Applied Genetics* 90:1133–1137.

Lutman, P. J. W. 1993. The occurrence and persistence of volunteer oilseed rape *(Brassica napus). Aspects of Applied Biology* 35:29–36.

Lutman, P. J. W. 1999. *Gene flow and agriculture—Relevance for transgenic crops.* Staffordshire: British Crop Protection Council.

Lynch, M., and J. S. Conery. 2000. The evolutionary fate and consequences of duplicate genes. *Science* 290:1151–1153.

Lynch, M., and B. Walsh. 1998. *Genetics and analysis of quantitative traits.* Sunderland: Sinauer.

Madsen, K. H., G. S. Poulsen, J. R. Fredshavn, J. E. Jansen, P. Steen, and J. C. Streibig. 1998. A method to study competitive ability of hybrids between seabeet *(Beta vulgaris* ssp. *maritima)* and glyphosate tolerant sugarbeet *(B. vulgaris* ssp. *vulgaris). Acta Agriculturae Scandinavica.* Section B, *Soil and Plant Science* 48:170–174.

Majewski, J., and F. M. Cohan. 1999. Adapt globally, act locally: The effect of selective sweeps on bacterial sequence diversity. *Genetics* 152:1459–1474.

Majumder, N. D., T. Ram, and A. C. Sharma. 1997. Cytological and morphological variation in hybrid swarms and introgressed population of interspecific hybrids *(Oryza rufipogon* Griff. × *Oryza sativa* L.) and its impact on evolution of intermediate types. *Euphytica* 94:295–302.

Mallory-Smith, C. A., J. Snyder, J. L. Hansen, Z. Wang, and R. S. Zemetra. 1999. Potential for gene flow between wheat *(Triticum aestivum)* and jointed goatgrass *(Aegilops cylindrica)* in the field. In *Gene flow and agriculture—Relevance for transgenic crops,* ed. P. J. W. Lutman, 165–169. Staffordshire: British Crop Protection Council.

Manasse, R. 1992. Ecological risks of transgenic plants: Effects of spatial dispersion of gene flow. *Ecological Applications* 2:431–438.

Manasse, R., and P. Kareiva. 1991. Quantifying the spread of recombinant genes and organisms. In *Assessing ecological risks of biotechnology,* ed. L. R. Ginzburg, 215–232. Stoneham: Butterworth-Heinemann.

Mange, A. P., and E. P. Mange. 1990. *Genetics: Human aspects.* Sunderland: Sinauer. 2nd ed.

Marchais, L. 1994. Wild pearl millet population *(Pennisetum glaucum,* Poaceae) integrity in agricultural Sahelian areas. An example from Keita (Niger). *Plant Systematics and Evolution* 189:233–245.

Marshall, D. R. 1977. The advantages and hazards of genetic homogeneity. *Annals of the New York Academy of Sciences* 287:1–20.

Martínez-Soriano, J. P. R., and D. S. Leal-Klevezas. 2000. Transgenic maize in Mexico: No need for concern. *Science* 287:1399.

Martinsen, G. D., T. G. Whitham, R. J. Turek, and P. Keim. 2001. Hybrid populations selectively filter gene introgression between species. *Evolution* 55:1325–1335.

Marvier, M. A. 2001. Ecology of transgenic crops. *American Scientist* 89:160–167.

Marvier, M. A., and P. Kareiva. 1999. Extrapolating from field experiments that remove herbivores to population-level effects of herbivore resistance transgenes. In *Ecological effects of pest resistance genes in managed ecosystems,* ed. P. L. Traynor and J. H. Westwood, 57–64. Blacksburg: Information Systems for Biotechnology.

Marvier, M. A., E. Meir, and P. M. Kareiva. 1999. How do the design of monitoring and control strategies affect the chance of detecting and containing transgenic weeds? In *Methods for risk assessment of transgenic plants.* Vol. 3, *Ecological risks and prospects of transgenic plants, where do we go from here? A dialogue between biotech industry and science,* ed. K. Ammann, Y. Jacot, G. Simonsen, and G. Kjellsson, 109–122. Basel: Birkhauser.

Maynard Smith, J., and J. Haigh. 1974. The hitch-hiking effects of a favorable gene. *Genetical Research* 23:23–25.

McFarlane, J. S. 1975. Naturally occurring hybrids between sugar beet and *Beta macrocarpa* in the Imperial Valley of California. *Journal of the American Society of Sugar Beet Technologists* 18:245–251.

McGloughlin, M. M. 2000. Biotech crops: Rely on the science. *Washington Post,* 14 June, p. A39.

McHughen, A. 2000. *Pandora's picnic basket.* Oxford: Oxford University Press.

McNaughton, I. H. 1995a. Swedes and rapes. In *Evolution of crop plants,* ed. J. Smartt and N. W. Simmonds, 68–75. Harlow: Longman. 2nd ed.

McNaughton, I. H. 1995b. Turnip and relatives. In *Evolution of crop plants,* ed. J. Smartt and N. W. Simmonds, 62–68. Harlow: Longman. 2nd ed.

McNeilly, T., and J. Antonovics. 1968. Evolution in closely adjacent populations. IV. Barriers to gene flow. *Heredity* 23:219–238.

McPartlan, H. C., and P. J. Dale. 1994. An assessment of gene transfer by pollen from field-grown transgenic potatoes to non-transgenic potatoes and related species. *Transgenic Research* 3:216–225.

Medina Filho, H. P., A. Carvalho, R. M. L. Ballve, R. Bordignon, M. M. A. de Lima, and L. C. Fazuoli. 1995. Isoenzymic evidence on the interspecific origin of Piāta coffee. *Bragantia* 54:263–273.

Mehlenbacher, S. A., V. Cociu, and L. F. Hough. 1990. Apricots *(Prunus).* In *Genetic resources of temperate fruit and nut crops,* vol. 1, ed. J. N. Moore and J. R. Ballington, Jr., 63–108. Wageningen: International Society for Horticultural Science.

Mehra, K. L. 1962. Natural hybridization between *Eleusine coracana* and *E. africana* in Uganda. *Journal of the Indian Botanical Society* 41:531–539.

Meinke, D. W., J. M. Cherry, C. Dean, S. D. Rounsley, and M. Koornneef. 1998. *Arabidopsis thaliana:* A model plant for genome analysis. *Science* 282:662, 679–682.

Mellon, M., and J. Rissler. 1998. *Now or never: Serious new plans to save a natural pest control.* Cambridge: Union of Concerned Scientists.

Metz, P. L. J., E. Jacobsen, J. P. Nap, A. Pereira, and W. J. Stiekema. 1997. The impact of biosafety of the phosphinothricin-tolerance transgene in inter-specific *B. rapa* times

B. napus hybrids and their successive backcrosses. *Theoretical and Applied Genetics* 95:442–450.

Meunier, J., and J. J. Hardon. 1976. Interspecific hybrids between *Elaeis guineensis* and *E. oleifera*. In *Oil palm research,* ed. R. H. V. Corley, J. J. Hardon, and B. J. Wood, 127–138. Amsterdam: Elsevier.

Meyn, O., and W. A. Emboden. 1987. Parameters and consequences of introgression in *Salvia apiana* × *S. mellifera* (Lamiaceae). *Systematic Botany* 12:390–399.

Mikkelsen, T. R., B. Andersen, and R. B. Jørgensen. 1996. The risk of crop transgene spread. *Nature* 380:31.

Mills, L. S., and F. W. Allendorf. 1996. The one-migrant-per-generation rule in conservation and management. *Conservation Biology* 10:1509–1518.

Milne, R. I., and R. J. Abbott. 2000. Origin and evolution of invasive naturalised material of *Rhododendron ponticum* L. in the British Isles. *Molecular Ecology* 9:541–556.

Mitchell, R. J., M. H. D. Auld, J. M. Hughes, and R. H. Marrs. 2000. Estimates of nutrient removal during heathland restoration on successional sites in Dorset, southern England. *Biological Conservation* 95:233–246.

Mok, D. W. S., and S. J. Peloquin. 1975. Three mechanisms of 2n pollen formation in diploid potatoes. *Canadian Journal of Genetics and Cytology* 17:217–225.

Montalvo, A. M., and N. C. Ellstrand. 2001. Non-local transplantation and outbreeding depression in the subshrub *Lotus scoparius* (Fabaceae). *American Journal of Botany* 88:258–269.

Morici, C. 1998. *Phoenix canariensis* in the wild. *Principes* 42(2): 85–89, 92–93.

Morishima, H., Y. Sano, and H. I. Oka. 1992. Evolutionary studies in rice and its wild relatives. *Oxford Surveys in Evolutionary Biology* 8:135–184.

Morris, W. F., P. M. Kareiva, and P. L. Raymer. 1994. Do barren zones and pollen traps reduce gene escape from transgenic crops? *Ecological Applications* 4:157–165.

Mossberg, B., L. Stenberg, and S. Ericsson. 1992. *Den nordiska floran.* Stockholm: Wahlström & Widstrand.

Mücher, T., P. Hesse, M. Pohl-Orf, N. C. Ellstrand, and D. Bartsch. 2000. Characterization of weed beet in Germany and Italy. *Journal of Sugar Beet Research* 37(3):19–38.

Mueller, U. G., and L. L. Wolfenbarger. 1999. AFLP genotyping and fingerprinting. *Trends in Ecology and Evolution* 14:389–394.

Muir, W. M., and R. D. Howard. 1999. Possible ecological risks of transgenic organism release when transgenes affect mating success: Sexual selection and the Trojan gene hypothesis. *Proceedings of the National Academy of Sciences, USA* 92:13853–13856.

Muir, W. M., and R. D. Howard. 2001. Fitness components and ecological risk of transgenic release: A model using Japanese medaka *(Oryzias latipes). American Naturalist* 158:1–16.

Nason, J. D., and N. C. Ellstrand. 1993. Estimating the frequencies of genetically distinct classes of individuals in hybridized populations. *Journal of Heredity* 84:1–12.

Nason, J. D., N. C. Ellstrand, and M. L. Arnold. 1992. Patterns of hybridization and introgression in populations of oaks, manzanitas, and irises. *American Journal of Botany* 79:101–111.

Nason, J. D., E. A. Herre, and J. L. Hamrick. 1998. The breeding structure of a tropical keystone plant resource. *Nature* 391:685–687.

Nassar, N. M. A. 1980. Attempts to hybridize wild *Manihot* species with cassava. *Economic Botany* 34:12–14.

Nassar, N. M. A. 1984. Natural hybridization between *Manihot reptans* Pax and *M. alutacea* Rogers & Appan. *Canadian Journal of Plant Science* 64:423–425.

Nassar, N. M. A., M. A. Vieira, C. Vierira, and D. Grattapaglila. 1998. Molecular and embryonic evidence of apomixis in cassava interspecific hybrids. *Canadian Journal of Plant Science* 78:349–352.

National Academy of Sciences. 1989. *Field testing genetically modified organisms: Framework for decisions.* Washington, D.C.: National Academy Press.

National Academy of Sciences. 2000. *Genetically modified pest-protected plants: Science and regulation.* Washington, D.C.: National Academy Press.

National Academy of Sciences. 2002. *Environmental effects of transgenic plants.* Washington, D.C.: National Academy Press.

Nei, M. 1978. Estimation of average heterozygosity and genetic distance from a small number of individuals. *Genetics* 89:583–590.

Neuffer, B., H. Auge, H. Mesch, U. Amarell, and R. Brandl. 1999. Spread of violets in polluted pine forests: Morphological and molecular evidence for the ecological importance of interspecific hybridization. *Molecular Ecology* 8:365–377.

Newstrom, L. E. 1991. Evidence for the origin of the cultivated chayote, *Sechium edule* (Cucurbitaceae). *Economic Botany* 45:410–428.

Ng, N. Q. 1995. Cowpea. In *Evolution of crop plants,* ed. J. Smartt and N. W. Simmonds, 326–332. Harlow: Longman. 2nd ed.

Nickson, T. J., and G. P. Head. 1999. Environmental monitoring of genetically modified crops. *Journal of Environmental Monitoring* 1:101N–105N.

Noor, M. A. F. 1999. Reinforcement and other consequences of sympatry. *Heredity* 83:503–508.

Novak, S. J., D. E. Soltis, and P. S. Soltis. 1991. Ownbey's Tragopogons: 40 years later. *American Journal of Botany* 78:1586–1600.

Nurminiemi, M., and O. A. Rognli. 1993. *Kulturplanter og risiko for gensprendning.* Institutt for bioteknologifag, Norges Landbrukshøgskole.

Oard, J., M. A. Cohn, S. Linscombe, D. Gealy, and K. Gravois. 2000. Field evaluation of seed production, shattering, and dormancy in hybrid populations of transgenic rice *(Oryza sativa)* and the weed, red rice *(Oryza sativa).* *Plant Science* 157:13–22.

Obrycki, J. J., J. E. Losey, O. R. Taylor, and L. C. H. Jesse. 2001. Transgenic insecticidal corn: Beyond insecticidal toxicity to ecological complexity. *BioScience* 65:353–361.

Ochoa, C. M. 1990. *The potatoes of South America: Bolivia.* Cambridge: Cambridge University Press.

O'Hanlon, P. C., R. Peakal, and D. T. Briese. 1999. Amplified fragment length polymorphism (AFLP) reveals introgression in weedy *Onopordum* thistles: Hybridization and invasion. *Molecular Ecology* 8:1239–1246.

Oka, H. I. 1957. Phylogenetic differentiation of cultivated rice. XV. Complementary lethal genes in rice. *Japanese Journal of Genetics* 32:83–87.

Oka, H. I. 1988. *Origin of cultivated rice.* Tokyo: JSSP/Elsevier.

Oka, H. I. 1992. Ecology of wild rice planted in Taiwan: II. Comparison of two populations with different genotypes. *Botanical Bulletin of Academia Sinica* 33:75–84.

Oka, H. I., and W. T. Chang. 1959. The impact of cultivation on populations of wild rice, *Oryza sativa* f. *spontanea.* *Phyton* 13:105–117.

Oka, H. I., and W. T. Chang. 1961. Hybrid swarms between wild and cultivated rice species, *Oryza perennis* and *O. sativa.* *Evolution* 15:418–430.

Oliver, M. J., J. E. Quisenberry, N. L. G. Trolinder, and D. L. Keim. 1998. *Control of plant gene expression.* United States Patent 5 723 765 assignee Delta and Pine Land Co.

Olmo, H. P. 1995. Grapes. In *Evolution of crop plants,* ed. J. Smartt and N. W. Simmonds, 485–490. Harlow: Longman. 2nd ed.

Olmo, H. P., and A. Koyama. 1980. Natural hybridization of indigenous *Vitis californica* and *V. girdiana* with cultivated *vinifera* in California. *Proceedings of the third international symposium on grape breeding,* 31–41. Davis: University of California.

Orozco-Castillo, C., K. J. Chalmers, R. Waugh, and W. Powell. 1994. Detection of genetic diversity and selective gene introgression in coffee using RAPD markers. *Theoretical and Applied Genetics* 87:934–940.

Ouazzani, N., R. Lumaret, P. Villemur, and F. Di Giusto. 1993. Leaf allozyme variation in cultivated and wild olive trees (*Olea europaea* L.). *Journal of Heredity* 84:34–42.

Ouborg, N. J., Y. Piquot, and J. M. Van Groenendael. 1999. Population genetics, molecular markers, and the study of dispersal in plants. *Journal of Ecology* 87:551–568.

Palmer, J. D., K. L. Adams, Y. Cho, C. L. Parkinson, Y.-L. Qiu, and K. Song. 2000. Dynamic evolution of plant mitochondrial genomes: Mobile genes and introns and highly variable mutation rates. In *Variation and evolution in plants and microorganisms: Toward a new synthesis fifty years after Stebbins,* ed. F. J. Ayala, W. M. Fitch, and M. T. Clegg, 35–57. Washington, D.C.: National Academy Press.

Panetsos, C. A., and H. G. Baker. 1967. The origin of variation in "wild" *Raphanus sativus* (Cruciferae) in California. *Genetica* 38:243–274.

Parker, C., and M. L. Dean. 1976. Control of wild rice in rice. *Pesticide Science* 7:403–416.

Parker, P. G., A. A. Snow, M. D. Schug, G. C. Booton, and P. A. Fuerst. 1998. What molecules can tell us about populations: Choosing and using a molecular marker. *Ecology* 79:361–382.

Paterson, A. H., K. F. Schertz, Y. R. Lin, S. C. Liu, and Y. L. Chang. 1995. The weediness of wild plants: Molecular analysis of genes influencing dispersal and persistence of johnsongrass, *Sorghum halepense* (L.) Pers. *Proceedings of the National Academy of Sciences, USA* 92:6127–6131.

Penner, G. A. 1996. RAPD analysis of plant genomes. In *Methods of genome analysis in plants,* ed. P. P. Jauhar, 251–268. Boca Raton: CRC Press.

Pessel, F. D., J. Lecomte, V. Emeriau, M. Krouti, A. Messean, and P. H. Gouyon. 2001. Persistence of oilseed rape (*Brassica napus* L.) outside of cultivated fields. *Theoretical and Applied Genetics* 102:841–846.

Phillips, S. M. 1972. A survey of the genus *Eleusine* Gaertn. (Gramineae) in Africa. *Kew Bulletin* 27:251–270.

Pohl-Orf, M., U. Brand, S. Driessen, P. R. Hesse, M. Lehnen, C. Morak, T. Mücher, C. Saeglitz, C. von Soosten, and D. Bartsch. 1999. Overwintering of genetically modified sugar beet, *Beta vulgaris* L. ssp. *vulgaris* as a source for dispersal of transgenic pollen. *Euphytica* 108:181–186.

Popova, G. 1923. Wild species of *Aegilops* and their mass-hybridisation with wheat in Turkestan. *Bulletin of Applied Botany* 13:475–482.

Powell-Abel, P., R. S. Nelson, B. De, N. Hoffmann, S. G. Rogers, R. T. Fraley, and R. N. Beachy. 1986. Delay of disease development in transgenic plants that express the tobacco mosaic virus CP gene. *Science* 232:738–743.

Prather, T. S., J. DiTomaso, and J. S. Holt. 2000. *Herbicide resistance: Definition and management strategies.* University of California, Division of Agriculture and Natural Resources, publication 8012. http://anrcatalog.ucdavis.edu/pdf/8012.pdf.

Pretty, J. 2001. The rapid emergence of genetic modification in world agriculture: Contested risks and benefits. *Environmental Conservation* 28:248–262.

Rabinowitz, D., C. R. Linder, R. Ortega, D. Begazo, H. Murguia, D. S. Douches, and C. F. Quiros. 1990. High levels of interspecific hybridization between *Solanum sparsipilum* and *S. stenotomum* in experimental plots in the Andes. *American Potato Journal* 67:73–81.

Raina, S. N., Y. Mukai, and M. Yamamoto. 1998. *In situ* hybridization identifies the diploid progenitor species of *Coffea arabica* (Rubiaceae). *Theoretical and Applied Genetics* 97:1204–1209.

Ravindran, P. N., K. N. Babu, B. Sasikumar, and K. S. Krishnamurthy. 2000. Botany and crop improvement of black pepper. In *Black pepper:* Piper nigrum, ed. P. N. Ravindran, 23–142. Amsterdam: Harwood Academic.

Rawal, K. M. 1975. Natural hybridization among wild, weedy, and cultivated *Vigna unguiculata* (L.) Walp. *Euphytica* 24:699–705.

Raybould, A. F., and A. J. Gray. 1993. Genetically modified crops and hybridization with wild relatives: A UK perspective. *Journal of Applied Ecology* 30:199–219.

Raybould, A. F., and A. J. Gray. 1994. Will hybrids of genetically modified crops invade natural communities? *Trends in Ecology and Evolution* 9:85–89.

Renno, J. F., T. Winkel, F. Bonnefous, and G. Benzançon. 1997. Experimental study of gene flow between wild and cultivated *Pennisetum glaucum. Canadian Journal of Botany* 75:925–931.

Rhymer, J. M., and D. Simberloff. 1996. Extinction by hybridization and introgression. *Annual Review of Ecology and Systematics* 27:83–109.

Ribeiro, J. M. C., and A. Spielman. 1986. The satyr effect: A model predicting parapatry and species extinction. *American Naturalist* 128:513–528.

Richards, A. J. 1997. *Plant breeding systems.* London: Chapman and Hall. 2nd ed.

Rick, C. M. 1987. Evolution of mating systems in cultivated plants. In *Plant evolutionary biology,* ed. L. D. Gottlieb and S. K. Jain, 133–147. New York: Chapman and Hall.

Rieger, M. A., T. D. Potter, C. Preston, and S. B. Powles. 2001. Hybridisation between *Brassica napus* L. and *Raphanus raphanistrum* L. under agronomic field conditions. *Theoretical and Applied Genetics* 103:555–560.

Rieseberg, L. H. 1991a. Homoploid reticulate evolution in *Helianthus:* Evidence from ribosomal genes. *American Journal of Botany* 78:1218–1237.

Rieseberg, L. H. 1991b. Hybridization in rare plants: Insights from case studies in *Cercocarpus* and *Helianthus*. In *Genetics and conservation of rare plants,* ed. D. A. Falk and K. E. Holsinger, 171–181. New York: Academic Press.

Rieseberg, L. H. 1995. The role of hybridization in evolution: Old wine in new skins. *American Journal of Botany* 82:944–953.

Rieseberg, L. H. 1997. Hybrid origins of plant species. *Annual Review of Ecology and Systematics* 28:359–389.

Rieseberg, L. H., and S. J. Brunsfeld. 1992. Molecular evidence and plant introgression. In *Molecular systematics in plants,* ed. P. E. Soltis, D. E. Soltis, and J. J. Doyle, 151–176. New York: Chapman and Hall.

Rieseberg, L. H., and S. E. Carney. 1998. Plant hybridization. *New Phytologist* 140 599–624.

Rieseberg, L. H., R. Carter, and S. Zona. 1990. Molecular tests of the hypothesized hybrid origin of two diploid *Helianthus* species. *Evolution* 44:1498–1511.

Rieseberg, L. H., and N. C. Ellstrand. 1993. What can molecular and morphological markers tell us about plant hybridization? *Critical Reviews in Plant Sciences* 12:213–241.

Rieseberg, L. H., M. J. Kim, and G. J. Seiler. 1999. Introgression between the cultivated sunflower and a sympatric wild relative, *Helianthus petiolaris* (Asteraceae). *International Journal of Plant Sciences* 160:102–108.

Rieseberg, L. H., and D. E. Soltis. 1991. Phylogenetic consequences of cytoplasmic gene flow in plants. *Evolutionary Trends in Plants* 5:65–84.

Rieseberg, L. H., and J. F. Wendel. 1993. Introgression and its consequences in plants. In *Hybrid zones and the evolutionary process,* ed. R. Harrison, 70–109. New York: Oxford University Press.

Rissler, J., and M. Mellon. 1996. *The ecological risks of engineered crops.* Cambridge: MIT Press.

Roach, B. T. 1995. Sugar canes. In *Evolution of crop plants,* ed. J. Smartt and N. W. Simmonds, 160–166. Harlow: Longman. 2nd ed.

Robbins, W. W., M. K. Bellue, and W. S. Ball. 1970. *Weeds of California.* Sacramento: California State Department of Agriculture.

Rogers, C. E., T. E. Thompson, and G. J. Seiler. 1982. *Sunflower species of the United States.* Bismarck: National Sunflower Association.

Rogers, D. J., and S. G. Appan. 1973. *Manihot, Manihotoides.* New York: Hafner Press.

Roseboro, K. 2001. "Gene-blocking" corn may help prevent GMO contamination. *Seed World* 139(1):18–19.

Rosellini, D., M. Pezzotti, and F. Veronesi. 2001 Characterization of transgenic male sterility in alfalfa. *Euphytica* 118:313–319.

Rosskopf, E. 1999. Report of the berry working group. In *Ecological effects of pest resistance genes in managed ecosystems,* ed. P. L. Traynor and J. H. Westwood, 67–72. Blacksburg: Information Systems for Biotechnology.

Roy, M., and R. Wu. 2001. Arginine decarboxylase transgene expression and analysis of environmental stress tolerance in transgenic rice. *Plant Science* 160:869–875.

Rubin, B. 1991. Herbicide resistance in weeds and crops, progress and prospects. In *Herbicide resistance in weeds and crops,* ed. J. C. Caseley, G. W. Cussans, and R. K. Atkin, 387–414. Oxford: Butterworth-Heinemann.

Rufener Al Mazyad, P., and K. Ammann. 1999. Biogeographical assay and natural gene flow. In *Methods for risk assessment of transgenic plants.*Vol. 3, *Ecological risks and prospects of transgenic plants,* ed. K. Ammann, Y. Jacot, V. Simonsen, and G. Kjellson, 95–98. Basel: Birkhauser.

Saeglitz, C., M. Pohl, and D. Bartsch. 2000. Monitoring gene flow from transgenic sugar beet using cytoplasmic male-sterile bait plants. *Molecular Ecology* 9:2035–2040.

Salick, J. 1992. Crop domestication and the evolution of cocona (*Solanum sessiliflorum* Dunal.). *Evolutionary Biology* 26:247–285.

Sánchez González, J. J., and J. A. Ruiz Corral. 1997. Teosinte distribution in Mexico. In *Gene flow among maize landraces, improved maize varieties and teosinte: Implications for transgenic maize,* ed. J. A. Serratos, M. C. Willcox, and F. Castillo González, 18–36. Mexico, DF: CIMMYT.

Santoni, S., and A. Bervillé. 1992. Evidence for gene exchanges between sugar beet (*Beta vulgaris* L.) and wild beets: Consequences for transgenic sugar beets. *Plant Molecular Biology* 20:578–580.

Sauer, J. D. 1995. Grain amaranths. In *Evolution of crop plants*, ed. J. Smartt and N. W. Simmonds, 8–10. Harlow: Longman. 2nd ed.

Saumitou-Laprade, P., G. J. A. Rouwendal, J. Cuguen, F. A. Krens, and G. Michaelis. 1993. Different CMS sources found in *Beta vulgaris* ssp. *maritima:* Mitochondrial variability in wild populations revealed by a rapid screening procedure. *Theoretical and Applied Genetics* 85:529–535.

Saxena, D., and G. Stotzky. 2001. Bt corn has a higher lignin content than non-Bt corn. *American Journal of Botany* 88:1704–1706.

Scientists' Working Group on Biosafety. 1998. *Manual for assessing ecological and human health effects of genetically engineered organisms*. Edmonds: Edmonds Institute.

Scott, S. E., and M. J. Wilkinson. 1999. Low probability of chloroplast movement from oilseed rape *(Brassica napus)* into wild *Brassica rapa*. *Nature Biotechnology* 17:390–393.

Second, G. 1982. Origin of the genic diversity of cultivated rice (*Oryza* spp.): Study of the polymorphism scored at forty isozyme loci. *Japanese Journal of Genetics* 57:25–57.

Seefeldt, S. S., R. Zemetra, F. L. Young, and S. S. Jones. 1998. Production of herbicide-resistant jointed goatgrass *(Aegilops cylindrica)* × wheat *(Triticum aestivum)* hybrids in the field by natural hybridization. *Weed Science* 46:632–634.

Senior, I. J., and P. J. Dale. 1999. Molecular aspects of multiple transgenes and gene flow to crops and wild relatives. In *Gene flow and agriculture—Relevance for transgenic crops*, ed. P. J. W. Lutman, 225–231. Staffordshire: British Crop Protection Council.

Sharma, D. R., R. Kaur, and K. Kumar. 1996. Embryo rescue in plants—A review. *Euphytica* 89:325–337.

Simmonds, N. W. 1962. *The evolution of the banana*. London: Longman.

Simmonds, N. W. 1981. *Principles of crop improvement*. London: Longman.

Simmonds, N. W. 1995. Potatoes. In *Evolution of crop plants*, ed. J. Smartt and N. W. Simmonds, 466–471. Harlow: Longman. 2nd ed.

Simpson, B. B., and M. C. Ogorzaly. 2001. *Economic botany: Plants in our world*. New York: McGraw-Hill. 3rd ed.

Singh, A. K. 1978. Cytogenetics of semi-arid plants. III. A natural interspecific hybrid of Cucubitaceae (*Citrullus colocynthis* Schrad. × *C. vulgaris* Schrad.). *Cytologia* 43:569–574.

Singh, A. K. 1995. Groundnut. In *Evolution of crop plants*, ed. J. Smartt and N. W. Simmonds, 246–250. Harlow: Longman. 2nd ed.

Singh, R. J., and T. Hymowitz. 1989. The genomic relationship among *Glycine soja* Sieb. and Zucc., *Glycine max* (L.) Merr. and *Glycine gracilis* Skvortz. *Plant Breeding* 103:171–173.

Singh, R. J., and T. Hymowitz. 1999. Soybean genetic resources and crop improvement. *Genome* 42:605–616.

Slatkin, M. 1987. Gene flow and the geographic structure of natural populations. *Science* 236:787–792.

Small, E. 1984. Hybridization in the domesticated-weed-wild complex. In *Plant biosystematics*, ed. W. F. Grant, 195–210. Toronto: Academic Press.

Smartt, J., and N. W. Simmonds. 1995. *Evolution of crop plants*. Harlow: Longman. 2nd ed.

Smith, S. E. 1989. Biparental inheritance of organelles and its implications for crop improvement. *Plant Breeding Reviews* 6:361–393.

Snow, A. A., B. Andersen, and R. B. Jørgensen. 1999. Costs of transgenic herbicide resistance introgressed from *Brassica napus* into weedy *B. rapa*. *Molecular Ecology* 8:605–616.

Snow, A. A., and P. Moran-Palma. 1997. Commercialization of transgenic plants: Potential ecological risks. *BioScience* 47:86–96.

Snow, A. A., P. Moran-Palma, L. H. Rieseberg, A. Wszelaki, and G. J. Seiler. 1998. Fecundity, phenology, and seed dormancy of F_1 wild-crop hybrids in sunflower (*Helianthus annuus*, Asteraceae). *American Journal of Botany* 85:794–801.

Snow, A. A., K. L. Uthus, and T. M. Culley. 2001. Fitness of hybrids between weedy and cultivated radish: Implications for weed evolution. *Ecological Applications* 11:934–943.

Snyder, J. R., C. A. Mallory-Smith, S. Balter, J. L. Hansen, and R. S. Zemetra. 2000. Seed production on *Triticum aestivum* by *Aegilops cylindrica* hybrids in the field. *Weed Science* 48:588–593.

Sobral, B. W. S., D. P. V. Braga, E. S. LaHood, and P. Keim. 1994. Phylogenetic analysis of chloroplast restriction enzyme site mutations in the *Saccharinae* Griesb. subtribe of the *Andropogoneae* Dumort. tribe. *Theoretical and Applied Genetics* 87:843–853.

Soltis, D. E., and P. S. Soltis. 1999. Polyploidy: Recurrent formation and genome evolution. *Trends in Ecology and Evolution* 14:348–352.

Soltis, P. S., and D. E. Soltis. 2000. The role of genetic and genomic attributes in the success of polyploids. In *Variation and evolution in plants and microorganisms: Toward a new synthesis fifty years after Stebbins*, ed. F. J. Ayala, W. M. Fitch, and M. T. Clegg, 310–329. Washington, D.C.: National Academy Press.

Spencer, L. J., and A. A. Snow. 2001. Fecundity of transgenic wild-crop hybrids of *Cucurbita pepo* (Cucurbitaceae): Implications for crop-to-wild gene flow. *Heredity* 86:694–702.

Spooner, D. M., and R. G. Van den Berg. 1992. Species limits and hypotheses of hybridization of *Solanum verthaultii* Hawkes and *S. tarijense* Hawkes: Morphological data. *Taxon* 41:685–700.

Stace, C. A. 1975a. Avena L. In *Hybridization and the flora of the British Isles*, ed. C. A. Stace, 573–574. London: Academic Press.

Stace, C. A. 1975b. *Hybridization and the flora of the British Isles*. London: Academic Press.

Stace, C. A. 1991. *New flora of the British Isles*. Cambridge: Cambridge University Press.

Staniland, B. K., P. B. E. McVetty, F. Lyle, S. Yarrow, G. Freyssinet, and M. Freyssinet. 2000. Effectiveness of border areas in confining the spread of transgenic *Brassica napus* pollen. *Canadian Journal of Plant Science* 80:521–526.

Stebbins, G. L. 1959. The role of hybridization in evolution. *Proceedings of the American Philosophical Society* 103:231–251.

Stebbins, G. L. 1969. The significance of hybridization for plant taxonomy and evolution. *Taxon* 18:26–35.

Stebbins, G. L. 1974. *Flowering plants: Evolution above the species level*. Cambridge: Belknap Press.

Stevenson, G. C. 1965. *Genetics and breeding of sugar cane*. London: Longmans.

Stewart, C. N. 1999. Insecticidal transgenes into nature: Gene flow ecological effects, relevancy, and monitoring. In *Gene flow and agriculture—Relevance for transgenic crops*, ed. P. J. W. Lutman, 179–189. Staffordshire: British Crop Protection Council.

Stewart, C. N. 2001. The utility of green fluorescent protein in transgenic plants. *Plant Cell Reports* 20:376–382.

Stewart, C. N., and C. S. Prakash. 1998. Chloroplast-transgenic plants are not a gene flow panacea. *Nature Biotechnology* 16:401.

Stotzky, G. 2002. Release, persistence, and biological activity in soil of insecticidal proteins from *Bacillus thuringiensis*. In *Genetically engineered organisms: Assessing environmental and human health effects,* ed. D. K. Letourneau and B. E. Burrows, 187–222. Boca Raton: CRC Press.

Suh, H. S., Y. I. Sato, and H. Morishima. 1997. Genetic characterization of weedy rice (*Oryza sativa* L.) based on morpho-physiology, isozymes and RAPD markers. *Theoretical and Applied Genetics* 94:316–321.

Sun, M., and H. Corke. 1992. Population genetics of colonizing success of weedy rye in northern California. *Theoretical and Applied Genetics* 83:321–329.

Suneson, C. A., Rachie, K. O., and G. S. Khush. 1969. A dynamic population of weedy rye. *Crop Science* 9:121–124.

Syvanen, M., and C. I. Kado. 2002. *Horizontal gene transfer.* San Diego: Academic Press. 2nd ed.

Talbert, L. E., J. F. Doebley, S. Larson, and V. L. Chandler. 1990. *Tripsacum andersonii* is a natural hybrid involving *Zea* and *Tripsacum:* Molecular evidence. *American Journal of Botany* 77:722–726.

Tang, L. H., and H. Morishima. 1988. Characteristics of weed rice strains. *Rice Genetics Newsletter* 5:70–72.

Templeton, A. R. 1986. Coadaptation and outbreeding depression. In *Conservation biology: The science of scarcity and diversity,* ed. M. Soulé, 105–116. Sunderland: Sinauer.

Thomas, H. 1995. Oats. In *Evolution of crop plants,* ed. J. Smartt and N. W. Simmonds, 132–137. Harlow: Longman. 2nd ed.

Thompson, J. D. 1991. The biology of an invasive plant. *BioScience* 41:393–401.

Thompson, T. E., and L. J. Grauke. 1990. Pecans and hickories *(Carya).* In *Genetic resources of temperate fruit and nut crops,* vol. 2, ed. J. N. Moore and J. R. Ballington, Jr., 837–904. Wageningen: International Society for Horticultural Science.

Tiedje, J. M., R. K. Colwell, Y. L. Grossman, R. E. Hodson, R. E. Lenski, R. N. Mack, and P. J. Regal. 1989. The planned introduction of genetically engineered organisms: Ecological considerations and recommendations. *Ecology* 70:298–315.

Till-Bottraud, I., X. Reboud, P. Brabant, M. Lefranc, B. Rherissi, F. Vedel, and H. Darmency. 1992. Outcrossing and hybridization in wild and cultivated foxtail millets: Consequences for the release of transgenic crops. *Theoretical and Applied Genetics* 83:940–946.

Tomiuk, J., T. P. Hauser, and R. Jørgensen. 2000. A- or C-chromosomes, does it matter for the transfer of transgenes from *Brassica napus? Theoretical and Applied Genetics* 100:750–754.

Tostain, S. 1992. Enzyme diversity in pearl millet (*Pennisetum glaucum* L.). 3. Wild millet. *Theoretical and Applied Genetics* 83:733–742.

Traynor, P. L., and J. H. Westwood. 1999. *Proceedings of a workshop on ecological effects of pest resistance genes in managed ecosystems.* Blacksburg: Information Systems for Biotechnology.

Turelli, M., N. H. Barton, and J. A. Coyne. 2001. Theory and speciation. *Trends in Ecology and Evolution* 16:330–343.

U.S. Congress, Office of Technology Assessment. 1993. *Harmful, non-indigenous species in the United States,* OTA-F-565. Washington, D.C.: U.S. Government Printing Office.

USDA (United States Department of Agriculture). 2001a. *Crop production 2000 summary.* Agricultural Statistics Board. USDA.

USDA (United States Department of Agriculture) 2001b. *Fruit and tree nut yearbook.* USDA Economic and Statistics System. USDA.

USDA (United States Department of Agriculture) 2001c. *Noncitrus fruits and nuts: Preliminary summary.* Agricultural Statistics Board. National Agricultural Statistics Survey. USDA.

USDA (United States Department of Agriculture) 2001d. *Vegetables 2000 summary.* USDA Economic and Statistics System. USDA.

Van Aken, J. 1999. *Centers of diversity: Global heritage of crop varieties threatened by genetic pollution.* Berlin: Greenpeace International.

Vanderborght, T. 1983. Evaluation of *Phaseolus vulgaris* wild and weedy forms. *Plant Genetic Resources Newsletter* 54:18–25.

Van der Maesen, L. J. G. 1986. *Cajanus* DC. and *Atylosia* W. & A. (Leguminosae). Wageningen Agricultural University Papers 85-4. Wageningen: University of Wageningen.

Van der Vossen, H. A. M. 1985. Coffee selection and breeding. In *Coffee: Botany, biochemistry, and production of beans and beverage,* ed. M. N. Clifford and K. C. Willson, 13–47. Westport: Avi Publishing.

Van Dijk, H., and P. Boudry. 1992. Genetic variability for life-histories in *Beta maritima.* In *International* Beta *genetic resources workshop,* ed. L. Frese, 4–16. Roma: International Board for Plant Genetic Resources.

Van Dijk, P., and J. Van Damme. 2000. Apomixis technology and the paradox of sex. *Trends in Plant Science* 5:81–84.

Van Raamsdonk, L. W. D., and L. J. G. Van der Maesen. 1996. Crop-weed complexes: The complex relationship between crop plants and their wild relatives. *Acta Botanica Nederlandica* 45:135–155.

Van Slageren, M. W. 1994. *Wild wheats: A monograph of* Aegilops *L. and* Amblyopyrum *(Jaub. & Spach) Eig (Poaceae).* Wageningen: Veenman Druckers.

Vavilov, N. I. 1929. *Studies on the origin of cultivated plants.* Leningrad: Institute of Applied Botany and Plant Breeding.

Viard, F., J. Bernard, and B. Desplanque. 2002. Crop-weed interactions in the *Beta vulgaris* complex at a local scale: Allelic diversity and gene flow within sugar beet fields. *Theoretical and Applied Genetics* 104:688–697.

Vigouroux, Y., H. Darmency, T. Gestat de Garambe, and M. Richard-Molard. 1999. Gene flow between sugar beet and weed beet. In *Gene flow and agriculture—Relevance for transgenic crops,* ed. P. J. W. Lutman, 83–88. Staffordshire: British Crop Protection Council.

Vilà, M., and C. M. D'Antonio. 1998. Hybrid vigor for clonal growth in *Carpobrotus* (Aizoaceae) in coastal California. *Ecological Applications* 8:1196–1205.

Von Bothmer, R., N. Jacobsen, C. Baden, R. B. Jørgensen, and I. Linde-Laursen. 1991. *An ecogeographical study of the genus* Hordeum. Rome: IBPGR.

Wagner, D. B., and R. W. Allard. 1991. Pollen migration in predominantly self-fertilizing plants: Barley. *Journal of Heredity* 82:302–304.

Wagner, W., D. Herbst, and S. H. Sohmer 1990. *Manual of the flowering plants of Hawai'i.* Honolulu: University of Hawaii Press.

Waines, J. G., and S. G. Hegde. 2002. Pollen mediated gene flow in bread wheat (*Triticum aestivum* L.) as affected by biological characteristics and environmental factors. *Crop Science.* In press.

Warwick, S. I., and L. D. Black. 1983. The biology of Canadian weeds. 61. Sorghum halepense (L.) Pers. *Canadian Journal of Plant Science* 63:997–1014.

Waser, N. M. 1993. Population structure, optimal outbreeding, and assortative mating in angiosperms. In *The natural history of inbreeding and outbreeding: Theoretical and empirical perspectives,* ed. N. W. Thornhill, 173–199. Chicago: University of Chicago Press.

Waser, N. M., and M. V. Price. 1985. Reciprocal transplants with *Delphinium nelsonii* (Ranunculaceae): Evidence for local adaptation. *American Journal of Botany* 72:1726–1732.

Weeden, N. F., and J. F. Wendel. 1989. Genetics of plant isozymes. In *Isozymes in plant biology,* ed. D. E. Soltis and P. S. Soltis, 46–72. Portland: Dioscorides.

Wendel, J. F. 1995. Cotton. In *Evolution of crop plants,* ed. J. Smartt and N. W. Simmonds, 358–366. Harlow: Longman. 2nd ed.

Wendel, J. F., C. L. Brubaker, and A. E. Percival. 1992. Genetic diversity in *Gossypium hirsutum* and the origin of upland cotton. *American Journal of Botany* 79:1291–1310.

Wendel, J. F., and R. G. Percy. 1990. Allozyme diversity and introgression in the Galapagos Islands endemic *Gossypium darwinii* and its relationship to continental *G. barbadense. Biochemical Ecology and Systematics* 18:517–528.

Wendel, J. F., R. Rowley, and J. McD. Stewart. 1994. Genetic diversity in and phylogenetic relationships of the Brazilian endemic cotton *Gossypium mustelinum. Plant Systematics and Evolution* 192:49–59.

Wendel, J. F., and N. F. Weeden. 1989. Visualization and interpretation of plant isozymes. In *Isozymes in plant biology,* ed. D. E. Soltis and P. S. Soltis, 5–45. Portland: Dioscorides.

Westman, A. L., B. M. Levy, G. J. Gilles, T. S. Spira, S. Rajapakse, D. W. Tonkyn, and A. G. Abbott. 2000. Application of AFLP markers to assess past and present gene escape from cultivated to wild strawberry species. Plant & Animal Genome VIII Conference. Abstract 416. http://www.intl-pag.org/pag/8/abstracts/pag8416.html.

Whitkus, R., J. Doebley, and J. F. Wendel. 1994. Nuclear DNA markers in systematics and evolution. In *DNA-based markers in plants,* ed. R. L. Philips and I. K. Vasil, 116–141. Dordrecht: Kluwer.

Whitton, J., D. E. Wolf, D. M. Arias, A. A. Snow, and L. H. Rieseberg. 1997. The persistence of cultivar alleles in wild populations of sunflowers five generations after hybridization. *Theoretical and Applied Genetics* 95:33–40.

Wijnheijmer, E., Brandenburg, W. A., and S. J. Ter Borg. 1989. Interactions between wild and cultivated carrots (*Daucus carota* L.) in the Netherlands. *Euphytica* 40:147–154.

Wilkes, H. G. 1977. Hybridization of maize and teosinte in Mexico and Guatemala and the improvement of maize. *Economic Botany* 31:254–293.

Wilson, E. O. 1992. *The diversity of life.* New York: Norton.

Wilson, H., and J. Manhart. 1993. Crop/weed gene flow: *Chenopodium quinoa* Willd. and *C. berlandieri* Moq. *Theoretical and Applied Genetics* 86:642–648.

Winicov, I. 2000. Alfin1 transcription factor overexpression enhances plant root growth under normal and saline conditions and improves salt tolerance in alfalfa. *Planta* 210:416–422.

Wipff, J. K., and C. Fricker. 2001. Gene flow from transgenic creeping bentgrass (*Agrostis stolonifera* L.) in the Willamette Valley, Oregon. *International Turfgrass Society Research Journal* 9:224–242.

Wójcicki, J. J., and K. Marhold. 1993. Variability, hybridization and distribution of *Prunus fruticosa* (Rosaceae) in the Czech Republic and Slovakia. *Polish Botanical Studies* 5:9–24.

Wolf, D. E., N. Takebayashi, and L. H. Rieseberg. 2001. Predicting the risk of extinction through hybridization. *Conservation Biology* 15:1039–1053.

Wolfe, A. D., and A. Liston, 1998. Contributions of PCR-based methods to plant systematics and evolutionary biology. In *Molecular systematics of plants*. Vol. 2, *DNA sequencing*, ed. D. E. Soltis, P. S. Soltis, and J. J. Doyle, 43–86. Boston: Kluwer Academic Publishers.

Wolfe, A. D., Q.-Y. Xiang, and S. R. Kephart. 1998. Assessing hybridization in natural populations of *Penstemon* (Scrophulariaceae) using hypervariable inter simple sequence repeat markers. *Molecular Ecology* 7:1107–1125.

Wolfenbarger, L. L., and P. R. Phifer. 2000. The ecological risks and benefits of genetically engineered plants. *Science* 290:2088–2093.

Wright, S. 1969. *Evolution and the genetics of populations*. Vol. 2, *The theory of gene frequencies*. Chicago: University of Chicago Press.

Wrigley, G. 1988. *Coffee*. Harlow: Longman.

Wrigley, G. 1995a. Coffee. In *Evolution of crop plants*, ed. J. Smartt and N. W. Simmonds, 438–443. Harlow: Longman. 2nd ed.

Wrigley, G. 1995b. Date palm. In *Evolution of crop plants*, ed. J. Smartt and N. W. Simmonds, 399–403. Harlow: Longman. 2nd ed.

Wycherley, P. R. 1995. Rubber. In *Evolution of crop plants*, ed. J. Smartt and N. W. Simmonds, 124–128. Harlow: Longman. 2nd ed.

Xu, J., R. W. Kerrigan, P. Callac, P. A. Horgen, and J. B. Anderson. 1997. Genetic structure of natural populations of *Agaricus bisporus*, the commercial mushroom. *Journal of Heredity* 88:482–88.

Zamir, D., and Y. Tadmor. 1986. Unequal segregation of nuclear genes in plants. *Botanical Gazette* 147:355–358.

Zemetra, R. S., J. Hansen, and C. A. Mallory-Smith. 1998. Potential for gene transfer between wheat *(Triticum aestivum)* and jointed goatgrass *(Aegilops cylindrica). Weed Science* 46:313–317.

Zohary, D. 1971. Origin of south-west Asiatic cereals: Wheats, barley, oats, and rye. In *Plant life of south-west Asia*, ed. P. H. Davis, P. C. Harper, and I. C. Hedge, 235–260. Edinburgh: Botanical Society of Edinburgh.

Zohary, D. 1995. Olive. In *Evolution of crop plants*, ed. J. Smartt and N. W. Simmonds, 379–382. Harlow: Longman. 2nd ed.

Zohary, D., and M. Hopf. 1993. *Domestication of plants in the Old World*. Oxford: Clarendon Press. 2nd ed.

Index

adaptation, 38–40, 42, 145, 147–49

Aegilops. See goatgrass

Afghanistan, 110

AFLPs (amplified fragment length polymorphisms), 66–67, 94, 100, 143

Africa: cassava in, 104; coffee in, 107–8, 115; cowpeas in, 109–10, 115; millet in, 90–91, 114, 129, 146–47, 154; oil palm in, 111; peanuts in, 98; rice in, 83–84, 114; sorghum in, 88–89; sugarcane in, 102

agamospermy, 23

Agaricus bisporus (button mushroom), 123

AgrEvo, 160

agriculture, industrial: and beets, 165, 168–69; and Continent-Island model, 42–43; and gene flow, 35–36; globally important crops in, 6–7, 76–81; and transgenic crops, 168–69, 173, 175

Agriculture and Biotechnology Strategies (Canada), Inc., 160

Agrostis (bentgrass), 122

alfalfa (*Medicago sativa*), 122, 126, 176; and *M. falcata*, 150

Allard, R. W., 10

alleles, 18, 148, 158; bolting, 69–72, 200; dominant, 182, 200–201; dominant vs. recessive, 174–75; and fitness, 144–46; in gene flow management, 198–99

allopolyploidy (amphiploidy), 23–24, 146, 148, 152; in coffee, 107–8; and gene flow management, 197–98; in peanuts, 98; in potatoes, 100, 102; in rapeseed, 93; in sorghum, 89; in sweet potatoes, 112; in wheat, 81

allozyme (isozyme) analysis, 17, 19, 63–65, 143, 154; in beets, 73–74, 163, 166; in cas-

sava, 104; in cotton, 92; in grapes, 116; in maize, 86; in millet, 91; in oats, 105; in rapeseed, 94–96; in rice, 84; in rye, 111; in sorghum, 89; in sunflowers, 99; in wheat, 82–83

almonds (*Amygdalus communis*), 122

Altieri, Miguel A., 171

amaranth, grain (*Amaranthus*), 122

Amaranthaceae, 18

Ammann, K., 125

amphiploidy. *See* allopolyploidy

Andersen, B., 94

Anderson, E., 10, 119

apomixis, 79, 196–97, 200

apples (*Malus*), 122, 129

apricots (*Prunus armeniaca*), 122

Arabidopsis thaliana, 176

Arachis hypogaea. See groundnuts

Arctostaphylos: A. coronipifolium, 37; *A. patula*, 19; *A. viscida*, 19

Arecaceae, 78

Argentina, 98

Argyranthemum fructescens, 37

Arizona, 21

Arnold, M. L., 25, 49

Asia: barley in, 88, 114; beans in, 97; chickpeas in, 109; cotton in, 92; elm in, 154; grapes in, 115; oil palm in, 111; peanuts in, 98; potatoes in, 100; rice in, 83, 150; soybeans in, 87, 114, 146; sugarcane in, 102; sweet potatoes in, 112; wheat in, 81

Asteraceae, 18, 78

Australasia, 103, 115

Australia, 20, 96, 105, 110

Austria, 83

Avena. See oats